Entropy and its Physical Meaning

Entropy and its Physical Meaning

J. S. DUGDALE
Emeritus Professor, University of Leeds

Taylor & Francis
Publishers since 1798

UK   Taylor & Francis ltd, 1 Gunpowder Square, London EC4A 3DE

USA   Taylor & Francis Inc., 325 Chestnut Street, 8th Floor, Philadelphia, PA 19106

Copyright © J. S. Dugdale 1996

Reprinted 1998

All rights reserved. No part of this publication may be reproduced, stored in a retrieval system, or transmitted, in any form or by any means, electronic, electrostatic, magnetic tape, mechanical, photocopying, recording or otherwise, without the prior permission of the copyright owner.

British Library Cataloguing in Publication Data

A catalogue record for this book is available from the British Library
ISBN 0-7484-0568-2 (cased)
ISBN 0-7484-0569-0 (paperback)

Library of Congress Cataloguing Publication Data are available

Cover design by Jim Wilkie

Typeset in Times 10/12pt by Keyset Composition, Colchester, Essex

Printed in Great Britain by T. J. International Ltd.

Contents

Contents

Contents

Preface

This book is a revised version of my earlier book *Entropy and Low Temperature Physics*, with here rather less emphasis on the low temperature aspects of the subject. The concept of entropy lies at the heart of thermodynamics and is often thought of as obscure, even mysterious. The aim of this book is thus the same as before, namely, to make accessible the idea of entropy and to encourage an intuitive appreciation of its nature and use. In this new version, I have added exercises for the reader; mostly they are meant to be straightforward tests of understanding but sometimes they are used to extend the coverage of the text.

I am very grateful to Bryan Coles for reading the manuscript, for his suggestions and for his encouragement. May I also express my thanks to Tony Guénault for valuable discussions, and the staff of Taylor & Francis for their help in preparing the manuscript for publication.

I am indebted to the Editor of *Annalen der Physik* for permission to reproduce Figure 10b.

Finally I wish to record my great debt to my teachers: at school Mr G. R. Noakes; as an undergraduate Dr C. Hurst; as a research student Sir Francis Simon.

Entropy in thermodynamics

1

Temperature and heat: some historical developments

Entropy, as we shall see, is defined in terms of temperature and heat, so we shall begin with a brief study of these quantities. We shall do this by having a look at what some of the great experimenters and thinkers on these topics have done and written.

Fahrenheit (1686–1736) produced the first thermometer which was accurately reproducible and this made possible a systematic quantitative study of temperature and heat. It seems to have been soon recognised that, when bodies at different temperatures were brought into contact and when all changes had ceased, a thermometer placed in contact with each body in turn gave the same reading.

This is a most important observation, so important indeed that it has since been given the status of a law of thermodynamics – it is often called the zeroth law since the first and second laws of thermodynamics had already been formulated before this fact of observation was 'canonised'. At the time, however, this 'equilibrium of heat' – or thermal equilibrium as we should now call it – gave rise to considerable confusion. For example, it was interpreted by some to mean that at equilibrium there was an equal amount of heat per unit volume throughout the different bodies. As we can now see, this interpretation arose from a failure to distinguish between heat and temperature.

Joseph Black (1728–1799) did much to clarify this and other questions relating to the nature of heat and temperature. His work was published posthumously (1803) in his *Lectures on the Elements of Chemistry* edited by his friend, who had also been a pupil and colleague, John Robison.[1] Black distinguished clearly between quantity and intensity of heat, that is, between heat and temperature. He also introduced the concepts of heat capacity and

---

[1]Superior figures refer to Notes on pp. 181–3.

latent heat and by his experiments laid the foundations of the science of calorimetry.

A quotation from Robison's notes which are appended to his edition of Black's *Lectures* reads 'Before the year 1765, Dr Black had made many experiments on the heats communicated to water by different solid bodies, and had completely established their regular and steady differences in this respect. . . .' Already, therefore, by 1765 Black had demonstrated that different substances had different heat capacities. His method of measuring them was by what we now call 'the method of mixtures'. Here is Robison's description of the method and of the precautions taken to ensure accuracy:

> Dr Black estimated the [heat] capacities, by mixing the two bodies in equal masses, but of different temperatures; and then stated *their capacities as inversely proportional to the changes of temperature of each by the mixture.* Thus, a pound of gold, of the temperature 150°, being suddenly mixed with a pound of water, of the temperature 50°, raises it to 55° nearly: therefore the capacity of gold is to that of an equal weight of water as 5 to 95, or as 1 to 19; for the gold loses 95°, and the water gains 5°. . . .
>
> These experiments require the most scrupulous attention to many cir-cumstances which may affect the result.
>
> 1 The mixture must be made in a very extended surface, that it may quickly attain the medium temperature.
> 2 The stuff which is poured into the other should have the temperature of the room, that no change may happen in the pouring it out of its containing vessel.
> 3 The effect of the vessel in which the mixture is made must be con-sidered.
> 4 Less chance of error will be incurred when the substances are of equal bulk.
> 5 The change of temperature of the mixture, during a few successive moments, must be observed, in order to obtain the real temperature at the begin-ning.
> 6 No substances should be mixed which produce any change of temperature by their chemical action, or which change their temperature, if mixed when of the same temperature.
> 7 Each substance must be compared in a variety of temperatures, lest the ratio of the capacities should be different in different temperatures.
>
> When all of these circumstances have been duly attended to, we obtain the measure of the *capacities* of different substances for heat.[2]

To this day calorimetry requires 'the most scrupulous attention to many circumstances which may affect the result'. A great deal of scientific effort still goes into the accurate measurement of heat capacities over a wide range of temperatures and indeed this is still one of the primary measurements in thermodynamics. From it is derived the great bulk of our information about the thermal properties of substances (in particular, as we shall see, their internal energy and entropy).

The nature of heat

We see therefore that calorimetry had developed into a quantitative science by the end of the eighteenth century. However, there was still a great division of opinion about the nature of heat. In 1783, for example, Henry Cavendish (1731–1810), writing about 'the cold generated by the melting of ice' and 'the heat produced by the freezing of water', makes the following observation in a footnote:

> I am informed that Dr Black explains the above mentioned phenomena in the same manner; only, instead of using the expression, heat is generated or produced, he says, latent heat is evolved or set free; but as this expression relates to an hypothesis depending on the supposition, that the heat of bodies is owing to their containing more or less of a substance called the matter of heat; and as I think Sir Isaac Newton's opinion, that heat consists in the internal motion of the particles of bodies, much the most probable, I chose to use the expression, heat is generated. . . .

In fact, it is probable that Black never held any theory of heat with great conviction since on the whole he seems to have felt that all theories were a waste of time. However, it does seem clear that while some scientists at that time thought of heat as 'a substance called the matter of heat' or as 'an igneous fluid' (later it was called caloric) others thought this an unnecessary hypothesis. (Incidentally, although Cavendish gives Newton the credit for having thought of heat as a form of motion, Newton was by no means the first to have this idea.)

At about this time, Count Rumford (Benjamin Thompson, 1753–1814) began his important experiments on the nature of heat. Here is an extract from his account of the experiments:[3]

> Being engaged lately, in superintending the boring of cannon, in the workshops of the military arsenal at Munich, I was struck with the very considerable degree of heat which a brass gun acquires, in a short time, in being bored; and with the still more intense heat (much greater than that of boiling water, as I found by experiment,) of the metallic chips separated from it by the borer.
>
> The more I meditated on these phenomena, the more they appeared to me to be curious and interesting. A thorough investigation of them seemed even to bid fair to give a farther insight into the hidden nature of heat; and to enable us to form some reasonable conjectures respecting the existence, or non-existence, of an *igneous fluid*: a subject on which the opinions of philosophers have, in all ages, been much divided. . . .
>
> From whence comes the heat actually produced in the mechanical operation above mentioned?

Rumford set out to answer this question by experiment. From a cannon he made a brass cylinder which would just fit over a blunt steel borer. This borer was forced against the bottom of the cylinder and the cylinder was made to turn on its axis by means of the boring machine driven by horses. Of the four experiments Rumford describes, one was concerned with the heat capacity of the metal chips produced in the boring process. This he showed was the same as that of an equal mass of the original metal. Of the other experiments, the third was the most striking. In this experiment Rumford surrounded the cylinder being bored by a wooden water-tight box. To continue in his own words:

> The box was filled with cold water (viz. at the temperature of 60°) and the machine was put in motion.
>
> The result of this beautiful experiment was very striking, and the pleasure it afforded me amply repaid me for all the trouble I had had, in contriving and arranging the complicated machinery used in making it.
>
> The cylinder, revolving at the rate of about 32 times in a minute, had been in motion but a short time, when I perceived, by putting my hand into the water, and touching the outside of the cylinder, that heat was generated; and it was not long before the water which surrounded the cylinder began to be sensibly warm.
>
> At the end of 1 hour I found, by plunging a thermometer into the water in the box, . . . that its temperature had been raised no less than 47 degrees. . . .
>
> At the end of 2 hours, reckoning from the beginning of the experiment, the temperature of the water was found to be raised to 178°.
>
> At 2 hours 20 minutes it was 200°; and at 2 hours 30 minutes it ACTUALLY BOILED!
>
> It would be difficult to describe the surprise and astonishment expressed in the countenances of the by-standers, on seeing so large a quantity of cold water heated, and actually made to boil, without any fire.

Rumford then computes the quantity of heat produced in the experiment: he estimates it to be equivalent to the heat needed to raise the temperature of 26.58 lb of water by 180°F.

> As the machinery used in this experiment could easily be carried around by the force of one horse, (though, to render the work lighter, two horses were actually employed in doing it,) these computations show further how large a quantity of heat might be produced, by proper mechanical contrivance, merely by the strength of a horse, without either fire, light, combustion, or chemical decomposition; and, in a case of necessity, the heat thus produced might be used in cooking victuals.

In summarising his conclusions from these experiments Rumford writes:

> And, in reasoning on this subject, we must not forget to consider that most

remarkable circumstance, that the source of the heat generated by friction, in these experiments, appeared evidently to be *inexhaustible*.

It is hardly necessary to add, that any thing which any *insulated* body, or system of bodies, can continue to furnish *without limitation*, cannot possibly be a *material substance*: and it appears to me to be extremely difficult, if not quite impossible, to form any distinct idea of any thing, capable of being excited, and communicated, in the manner the heat was excited and communicated in these experiments, except it be MOTION.

These experiments clearly showed a close connection between heat and work. However, they are only *qualitative*, although, as Joule later remarked, the data do in fact contain enough information to yield rough *quantitative* results. Mayer (1814–1878), a German physician and physicist, understood clearly the relationship between heat and work and was able to make a numerical estimate of the conversion factor. He is now recognised, along with Joule, as the discoverer of the first law of thermodynamics, but at the time his work received little recognition.

The finally decisive, quantitative studies on heat and work were those made by Joule (1818–1889) in the 1840s. Almost fifty years after Rumford's work Joule presented a paper to the Royal Society on this same subject entitled 'The Mechanical Equivalent of Heat'.[4] He writes:

> For a long time it had been a favourite hypothesis that heat consists of 'a force or power belonging to bodies', but it was reserved for Count Rumford to make the first experiments decidedly in favour of that view. That justly celebrated natural philosopher demonstrated by his ingenious experiments that the very great quantity of heat excited by the boring of cannon could not be ascribed to a change taking place in the calorific capacity of the metal; and he therefore concluded that the motion of the borer was communicated to the particles of metal, thus producing the phenomena of heat.

Joule then points out that if, in Rumford's third experiment, you assume the rate of working to be about one horsepower (as Rumford indicates) you can estimate that the work required to raise the temperature of 1 pound of water by 1°F is about 1000 foot-pounds which, as he observes, is not very different from the valule derived from his own experiments, namely 772 foot-pounds.

> In 1843 I announced the fact that 'heat is evolved by the passaage of water through narrow tubes', and that each degree of heat per lb of water required for its evolution in this way a mechanical force represented by 770 foot-pounds. Subsequently, in 1845 and 1847, I employed a paddle-wheel to produce the fluid friction, and obtained the equivalents 781.5, 782.1 and 787.6, respectively from the agitation of water, sperm-oil and mercury. Results so closely coinciding with one another, and with those previously derived from experiments with elastic fluids and the electromagnetic machine, left no doubt on my mind as to the existence of an equivalent relation between force and heat; but still it appeared

of the highest importance to obtain that relation with still greater accuracy. This I have attempted in the present paper.

Joule now goes on to describe the apparatus with which he carried out his experiments to determine 'the mechanical equivalent of heat'. The apparatus consisted essentially of a paddle-wheel with eight sets of revolving arms working between four sets of stationary vanes. The purpose of the paddle-wheel was to stir a liquid contained in the space between the vanes. The paddle-wheel, attached to a 'roller', was made to rotate by means of two lead weights suitably connected to the roller by pulleys and fine twine. We can follow the course of experiment best in Joule's own words:

> The method of experimenting was simply as follows: The temperature of the frictional apparatus having been ascertained and the weights wound up with the assistance of the stand . . . the roller was refixed to the axis. The precise height of the weights above the ground having then been determined by means of the graduated slips of wood, . . ., the roller was set at liberty and allowed to revolve until the weights reached the flagged floor of the laboratory, after accomplishing a fall of about 63 inches. The roller was then removed to the stand, the weights wound up again, and the friction renewed. After this had been repeated twenty times, the experiment was concluded with another observation of the temperature of the apparatus. The mean temperature of the laboratory was determined by observations made at the commencement, middle and termination of each experiment.
>
> Previously to, or immediately after each of the experiments, I made trial of the effect of radiation and conduction of heat to or from the atmosphere, in depressing or raising the temperature of the frictional apparatus. In these trials, the position of the apparatus, the quantity of water contained by it, the time occupied, the method of observing the thermometers, the position of the experimenter, in short everything, with the exception of the apparatus being at rest, was the same as in the experiments in which the effect of friction was observed.

Joule then describes in detail the results of these extremely careful experiments. At the end of the paper he writes:

> I will therefore conclude by considering it as demonstrated by the experiments contained in this paper. . . .
>
> 1st. That the quantity of heat produced by the friction of bodies, whether solid or liquid, is always proportional to the quantity of force expended. And
>
> 2nd. That the quantity of heat capable of increasing the temperature of a pound of water (weighed *in vacuo*, and taken at between 55° and 60°) by 1°Fahrenheit requires for its evolution the expenditure of a mechanical force represented by the fall of 772 lb through the space of one foot.

This brief account gives some idea of how the subject of heat had developed up to about the middle of the nineteenth century. As we shall see later, a great deal of work was going on at this time on a different aspect of the subject, that relating to the second law of thermodynamics. Before going on to this, however, I would like to look at the concepts of temperature and of heat from a different point of view.

2

Temperature and heat: a different approach

Once we recognise that thermodynamics and mechanics are intimately related, it is then logical to begin with quantities already familiar and well defined from such branches of physics as mechanics and electromagnetism, and to use them to define, with the help of experimental information, the specifically thermodynamic quantities such as temperature, internal energy and entropy.

Such 'mechanical' properties, derived from other branches of physics or chemistry, are pressure, volume, chemical composition, magnetisation, dielectric constant, refractive index and so on. In simple systems, such as are usually dealt with in physics, only a few of these variables need to be specified in order to describe the thermodynamic state of the system. For example, suppose that we have one gram of helium gas and that we fix its volume and pressure. Suppose that we now measure some property of the gas such as its viscosity, refractive index, heat conductivity or dielectric constant. We then find that provided the volume and the pressure of this one gram of helium are always the same, these other properties are also always the same no matter by whom they are measured or when or where. We may thus think of the mass of helium, its pressure and volume as the thermodynamic 'co-ordinates' of the system. It is characteristic of *thermodynamics* that it deals only with macroscopic, large-scale quantities of this kind and not with the variables which characterise individual atoms or molecules. A complete atomic description of the helium gas would specify the mass of the helium atoms, their momentum or kinetic energy, their positions and mutual potential energy. The atomic co-ordinates of one gram-atom of helium would number about 10^{24} parameters; by contrast the thermodynamic co-ordinates number only three (the mass of gas, its pressure and volume). Later on we shall see how these two very different descriptions of the gas are linked together.

It is clear that only under certain conditions do the mass, pressure and volume of the gas suffice to describe its state. For example, if a gas suddenly explodes from a container, the system is far too complicated to be described simply by pressure and volume, which are, in any case, no longer well enough defined to be very useful parameters. For a simple thermodynamic description (in terms of, say, the pressure and volume) to be adequate it is evident that the pressure must be uniform and the volume well defined. This restriction can be relaxed if the systems can be divided into a set of sub-systems in each of which these conditions are satisfied. At present, however, we shall assume that our systems are everywhere uniform.

The concept of thermodynamic equilibrium can be defined in terms of the thermodynamic co-ordinates of the system. If these co-ordinates (e.g. pressure, volume and mass for a simple system) do not change with time, and provided there are no large-scale flow processes going on, then the system is said to be in thermodynamic equilibrium.

In most of our discussions we shall assume that the mass and chemical composition of the systems we deal with remain unchanged. Nonetheless there are important examples in physics wherein the system is 'open', i.e. the mass is not fixed, for example, where electrons can flow from one conductor to another. Moreover, although it is outside the scope of this book, we must bear in mind that thermodynamics has had some of its greatest triumphs and most fruitful applications in dealing with chemical reactions.

Born[5] has shown how with the help of these 'mechanical' co-ordinates (whose definition comes entirely from other branches of physics) it is possible to define the essentially thermal concepts of 'insulating' and 'conducting'. Basically the idea is simple enough: if an insulating wall separates two systems the thermodynamic co-ordinates of one can be altered at will without influencing those of the other. If the wall between them is conducting, this is no longer true. (If we knew all about the concept of temperature, we would describe these situations by saying: if the systems are separated by an insulating wall, the temperature of one does not influence the other, whereas if they are linked by a conducting wall, it does. For our present purposes, however, we wish to describe these situation without explicitly using the concept or the word 'temperature'.)

The notions of insulating and conducting walls are, of course, idealisations although a silvered Dewar vessel with a vacuum space between the walls provides a good approximation to the former (especially at low temperatures) and a thin pure copper sheet to the latter. Even though the full detailed development of the argument outlined here is rather cumbersome, the main thing is to recognise that nevertheless it *is* possible to find satisfactory definitions of 'insulating' and 'conducting' walls without using the words 'heat' or 'temperature'.

The idea of an insulating wall then leads to the important notion of an 'adiabatic' process: if a system is entirely surrounded by an insulating wall the system is an adiabatic system and any processes that it can undergo are adiabatic processes.

The zeroth law of thermodynamics

Having in this way introduced the concepts of 'conducting' and 'insulating' walls we can go on to define thermal equilibrium. Consider two systems surrounded by insulating walls except where they have a common conducting wall. Then, whatever states the systems are in when they are first brought into contact, their thermodynamic co-ordinates will, if the systems are left to themselves, eventually stop changing. The systems are then said to be in *thermal equilibrium*.

The zeroth law of thermodynamics may then be stated as follows: two systems which are in thermal equilibrium with a third are in thermal equilibrium with each other.

This law is the basis for the concept of temperature. Any convenient body, whose state can readily be varied and which has a convenient thermodynamic co-ordinate (e.g. the length of a mercury column etc.), can be used to measure temperature, i.e. be a thermometer. If the thermodynamic state of the thermometer, after having come into thermal equilibrium with one system, A, is unchanged when brought into thermal contact with a second system, B, then A and B are, according to the zeroth law, in thermal equilibrium, and are said to have the same temperature. It is found that in the simplest thermometer only a single co-ordinate has to be measured in order to indicate its thermodynamic state. A measure of this co-ordinate is then a measure of the temperature on this particular temperature scale.

To avoid a total arbitrariness in the temperature scale we shall for the present use the gas scale of temperature. We can do this conveniently in one of two ways. In the constant volume gas thermometer we define the temperature θ_V by the relationship:

$$\theta_V \propto p \tag{1}$$

where p is the pressure at this temperature of a fixed mass of gas kept at constant volume, V.

On the constant pressure gas scale, we define the temperature θ_p by the relationship:

$$\theta_p \propto V \tag{2}$$

where now V is the volume of a fixed mass of gas under constant pressure at the temperature θ_p. So far these definitions allow us to measure only *ratios* of temperatures; the size of the degree is fixed by defining the normal melting point of ice as having the value $273.15°$ (cf. p. 36).

If the measurements of θ_V and θ_p are made with any gas in the limit of very low pressures, the two temperature scales are found to be the same

13

and, in addition, are found to be independent of the particular gas used. Because this temperature scale is independent of the particular gas chosen we shall make use of it until we are able to define a truly 'absolute' temperature scale. It is important to remember that the gas scale of temperature is an experimental one; temperatures on it are found by making measurements on real gases at low pressures. Temperatures on this gas scale will be denoted by θ and we shall refer to the scale as the 'gas scale' of temperature.

Equations of state

As we have already seen, only a few macroscopic parameters are needed to specify the thermodynamic state of a system: in the simplest case the mass, the pressure and the volume of the system. If we fix the mass of the system and measure its temperature as a function of pressure and volume we obtain in this way a relationship between p, V and θ. (Note: θ here is measured on the gas scale of temperature.) This relationship is called an equation of state. An example of such an equation of state is one due to Dieterici which describes quite well the behaviour of actual gases at moderate pressures. It is as follows:

$$p = \left[\frac{R\theta}{(V-b)} \right] e^{-a/R\theta V} \tag{3}$$

where R per mole of gas is a constant for *all* gases, and a and b are constants for a particular gas.

Although such equations strictly apply only to systems in thermodynamic equilibrium, nevertheless they can be used to calculate the work done when the system changes its state, provided that in these changes the system is always very close to equilibrium. Such changes in which the system really passes through a succession of equilibrium states are called *quasi-static* processes.

The work done in thermodynamic processes: quasi-static changes

I have emphasised that the fundamental measurements in thermodynamics are taken directly from mechanics or other branches of physics. In this way, for example, we know that in a quasi-static change of volume, the work done *on* the system is

$$W = -\int_{V_1}^{V_2} p \, dV \tag{4}$$

where V_2 is the volume of the system at the end of the change and V_1 that at the beginning. (The work done *by* the system would have the opposite sign.) p is, of course, the pressure at any instant during the change. Expressions can also be derived for the work done in other quasi-static changes, such as, for example, the magnetisation of a body in a magnetic field. Since, however, these expressions have to be defined and used with considerable care, and since the details are irrelevant to the main argument, I will not go into these questions here.[6]

In the expression (4) for the work done in a quasi-static change of volume two parameters are involved. One of these, the pressure, does not depend on the mass of the system; the other one, V, is (other things being equal) directly proportional to the mass. For this reason, the pressure is often referred to as the 'intensive' and the volume as the 'extensive' variable. A similar distinction is possible in other expressions for the work done in quasi-static processes.

The work done in quasi-static processes can always be represented graphically on a suitable diagram of state. For example, if on a pV diagram the pressure of a fluid is plotted against its volume for some particular path, the work done *by* the fluid for that path is represented by the area under the curve, since this area is equal to

$$\int_{V_1}^{V_2} p\, \mathrm{d}V$$

Since there are an unlimited number of paths having the same starting point, A, and the same end point, B, and since under each of these paths the area is in general different, the work done in taking the fluid from state A to state B depends not only on the initial and final states of the system, but on the particular path between them.

Work in irreversible processes

Before discussing irreversible processes we must first define what we mean by a reversible process. The test of a reversible process is suggested by its name but some care is needed in applying this test. We call a change in a system 'reversible' if we can restore the system to its original state and at the same time restore its environment to its original condition. By 'environment' I mean the apparatus or equipment, or anything else, *outside* the system which is affected by the change we are discussing. The important point is this: after *any* change you can always restore a given limited system to its original condition but usually you cannot do so without causing a permanent change in its environment. To sum up then: the test of whether a change is reversible or not is to try to restore the *status quo* everywhere. If you can do this, the change is reversible, otherwise not. A reversible change is thus

15

an idealisation which can never be fully realised in practice. Nevertheless, the concept of reversibility is a vital one in thermodynamics and we shall meet it again in connection with the second law of thermodynamics.

An example of a simple reversible process is the slow compression of a gas in an insulated cylinder by a piston which has negligible friction with the cylinder walls. If the force on the piston exceeds that exerted by the pressure of the gas by an infinitesimal amount, the piston moves inwards and compresses the gas. If the force on the piston is reduced so as to be infinitesimally less than that exerted by the pressure of the gas the piston moves outwards and the gas expands towards its original volume.

Now we shall contrast this with two examples of irreversible processes and indicate how you can calculate the work done in such processes. If, in the reversible process just described, the piston had appreciable friction with the cylinder walls, the process would no longer be reversible. On the compression stroke more work would be required than if there were no friction. On the return stroke, however, this work is not recovered, but rather an additional amount of energy is again expended in overcoming friction. We therefore have an irreversible process. The work done against the frictional force, F (assumed constant), is just the product of F and the distance moved by the piston.

Here is another example. Suppose that you stir a fluid in a container with a paddle-wheel. To move the wheel at all you must exert a couple on it, and from the size of the couple and the number of revolutions of the wheel you can calculate the work done in the process. If the fluid is in a thermally insulated container the temperature of the fluid rises because of the stirring. If you reverse the direction of stirring the temperature of the fluid does not fall and you do not recover the work done.

We have now shown how it is possible to measure work in a number of processes and our next step is to consider the relationship between work, heat and internal energy, which are related by the first law of thermodynamics.

Exercises

Q1 Give an example to illustrate the difference between a system *in thermodynamic equilibrium* and a system *in a steady state*.

Q2 Express as partial differential coefficients the volume thermal expansion coefficient, the isothermal bulk modulus, the isothermal compressibility and the isothermal magnetic susceptibility.

Q3 $10^{-3} \, \text{m}^3$ of lead is compressed reversibly and isothermally at room temperature from 1 to 1000 atmospheres pressure. How much work is done on the lead? The isothermal compressibility of lead, $\beta_T = -V^{-1}(\partial V/\partial p)_T$, is assumed independent of pressure with the value of 2.2×10^{-6} atm^{-1}. Take 1 atm = 10^5 Pa.

Q4 Water flows at constant temperature through a pipe at a rate of 1 m^3 s^{-1}. The input pressure is 2×10^6 Pa and the output pressure is one half this. How much work is done in 1000 s?

3

The first law of thermodynamics

The mechanical theory of heat, as it is sometimes called, was first firmly established by the very careful experiments of Joule which we read about earlier. In these experiments the essential features were as follows. Work was done on a known mass of water (or other fluid), thus raising its temperature; the change in temperature was then measured very accurately. Joule ensured that as far as possible no heat entered or left the water so that the temperature change was entirely a result of the work. By performing the work in a variety of different ways (for example, electrically, by stirring or by friction) he was able to show that to change the temperature of the same mass of water through the same temperature interval always required the *same* amount of work no matter by what method the work was done. He found similar results with other liquids.

We interpret these experiments and generalise from them as follows: if a thermally isolated system is taken from state A to state B, the work required depends only on the states A and B and not on the intervening states or the method of doing the work.

This statement allows us to define a new thermodynamic co-ordinate, called the internal energy function, U, with which to characterise the state of a thermodynamic system. If we choose some convenient state A as the reference state of a specified thermodynamic system, then the work done in taking it by adiabatic processes to any other state B depends on B and not on the path. If this work is W we can thus write:

$$U_B - U_A = W \tag{5}$$

In this way we *define* the difference in internal energy between B and A as equal to the work done in going from A to B by adiabatic means. It is because W is independent of the path (for a thermally isolated system) that U_B has a unique value for each state of the system, provided of course that the same reference state is always used. (It is to be noted that it is *not* always possible to go from state A to state B by adiabatic processes alone; it is, however, always possible to go either from A to B or from B to A by such processes and this is sufficient to define U.) We now imagine that by suitable measurements of the work done in adiabatic processes a value of U is associated with each possible state of the system of interest. If we now relax the condition that the changes are to be adiabatic, we shall then generally find that in going from state A to state B

$$\Delta U = U_B - U_A \neq W$$

We now define the difference between ΔU and W (this difference is zero in adiabatic changes) as a measure of the heat Q which has entered the system in the change. We shall treat the heat *entering* a system as *positive*. Thus

$$Q = \Delta U - W$$

or

$$\Delta U = Q + W \qquad\qquad\qquad\qquad (6)$$

which is a statement of the first law of thermodynamics. Notice that here heat *entering* the system and work done *on* the system are regarded as positive.

To recapitulate this important argument: it is found experimentally that in a wide variety of processes the work done in taking a thermally isolated system from one state to another depends only on the two states and *not* on the path or the method of doing the work. We can thus associate with each state of a system a quantity U, called the internal energy function, whose value for any given state of a system is measured by the work done on the system in bringing it to that state from a given reference state under adiabatic conditions. We now imagine that by these means we measure the internal energy relative to some reference state for all states of the system. In any arbitrary process (no longer necessarily adiabatic) the difference between the change in internal energy, ΔU, and the work done on the system is now defined as the heat which has entered the system during the process.

A number of points concerning the first law of thermodynamics need comment.

(i) Since we have no independent definition of Q, the quantity of heat involved in a thermodynamic process, it might appear that the first law of thermodynamics is a mere definition without physical content. The essence of the law, however, lies in its recognition that in adiabatic processes W depends only on the initial and final states of the system; this forms the basis for the definition of U, the internal energy function. This result concerning W is certainly based on experiment although it is applied in circumstances far beyond those for which it has been directly tested. The ultimate test of the first law, however, lies in the testing of its predictions; these have been verified over and over again.

(ii)   It is not at once evident that the quantity Q defined in equation (6) is the same as that determined in conventional heat capacity measurements. A detailed discussion of such methods in the light of equation (6) shows, however, that this is in fact so.

(iii) Let me emphasise again that in the approach outlined here (due originally to Born) the thermodynamic concept of quantity of heat has been introduced and defined in terms of purely mechanical quantities.

(iv)   Notice, too, that the idea of heat arises only when the state of a system changes. It is meaningless to talk about quantity of heat in a body. Quantity of heat is *not* a function of the state of a body like temperature, pressure, volume or mass: heat is a form of energy *in transit* and cannot be said to exist except when changes of state are occurring. In this respect it is of course similar to the thermodynamic concept of work which also is always associated with changes of state. It is equally meaningless to talk about the quantity of work in a body.

(v)   Note that the thermodynamic concept of heat conflicts, in some important ways, with the common usage of the word. We say that heat is transferred from one body to another and that, for example, the heat lost by one body is gained by the other. There is thus a strong implication that the heat was originally inside the first body and ended up after the transfer inside the other body. It is precisely this notion that we have to get rid of in order to think clearly about heat, work and internal energy.

(vi)   Equation (6) applies to any kind of change, reversible or irreversible. In order that U be definable the initial and final states *must* be states of equilibrium, but the intermediate states may be far from it.

The internal energy – an analogy

The following analogy may help to make clear the significance of the internal energy function. Let us suppose that a man has a bank account in which a certain sum of money is credited to him. He may add to this sum either by depositing cash at the bank or by paying into his account cheques from other people. Likewise he may diminish the amount of money in the account by drawing out cash or by issuing cheques payable from the account. The point is that the change in the amount of money in the account in any period of time is the algebraic sum of all cash paid in or taken out and of all the cheques paid into or drawn on the account. The *total* (shown in his account book) does not distinguish between cheque payments and cash payments; only the sum of the two counts. So it is with the internal energy function: heat or work do not contribute separately; only the sum of the two matters.

21

This analogy also makes clear another point. When the bank account is static and not changing, there is no need to talk of cash or cheques. It is only for the process of *transferring* money that we need say how the transfer is to be made, i.e. by cash or cheque. So it is in thermodynamics. As long as there is no exchange of energy with other bodies, the internal energy function of a body is fixed and specifies the energy state of the body completely. The use of the words 'heat' or 'work' in these circumstances is entirely inappropriate. However, as soon as there is any change of energy it is then important to specify if this is by heat or work. In brief, therefore, heat refers to a particular kind of energy *transfer*.

Atomic interpretation of internal energy

To see more vividly just what the internal energy of a body is, we need to change our point of view and look at the body on the atomic scale. At this stage it is simplest and not seriously misleading to describe the atoms in classical, as opposed to quantum, terms. In these terms we should expect to find, at any finite temperature of the body, that its atoms were moving about with velocities whose magnitude and distribution depend on the temperature of the body. We expect, in fact, that the higher the temperature the higher the mean square velocity of the atoms and the greater the spread of velocities about the mean. This is a well-known result from the kinetic theory of gases; it is equally true of solids and liquids at high temperatures. In solids, of course, the atoms tend to be associated with definite positions in space and in perfect crystals the mean positions of the atoms form a regular three-dimensional array. In addition to the kinetic energy of the atoms, which depends on the temperature, the atoms also possess potential energy from their mechanical interaction with the other atoms. This potential energy too will usually depend on the temperature of the body and also on its volume. The internal energy of the body can now be interpreted as the total of the kinetic and potential energy of all the atoms that compose it. In this way, we can interpret a large-scale thermodynamic quantity, the internal energy, whose changes are measured by calorimetry, in terms of the mechanical properties of atoms.

Classical thermodynamics has no need of this atomic interpretation; it forms a consistent and self-contained discipline in which all the concepts pertain to the large scale. On the other hand, if the internal energy of a body is just the total mechanical energy of the atoms in it and if this energy can be changed by doing work on the body or by heating it, then the first law of thermodynamics is just a manifestation of the law of conservation of energy in mechanics. The conversion of work into internal energy is then to be thought of as the change of large-scale directed mechanical motion (in, for example, the piston during the compression of a gas or in the paddle-wheel during the stirring of a fluid) into the small-scale random movements of atoms. As we have already implied, the kinetic theory of gases is built up on the basis of these ideas; its success in relating the large-scale properties of gases to the mechanics of atoms is unquestioned. We shall come back to the atomic point of view later when we try to understand entropy.

The internal energy as a function of temperature and volume

We know from the definition of heat that

$$\Delta U = Q + W \qquad (6)$$

Now apply this to a quasi-static infinitesimal change in which the work done is by a change of volume dV at pressure p. Let us denote the small quantity of heat entering the system by $đq$ so that we have:

$$dU = đq - p\,dV \qquad (7)$$

(The symbol $đq$ is used to denote an infinitesimal quantity which is not the differential[7] of a function.)

If the change takes place at constant volume

$$dU = đq$$

If this change causes a temperature rise $d\theta$, then the heat capacity at constant volume must be given by:

$$C_V = \frac{đq}{d\theta} = \left(\frac{\partial U}{\partial \theta}\right)_V \qquad (8)$$

Consequently, if we integrate this expression with V constant

$$U_2 - U_1 = \int_{\theta_1}^{\theta_2} C_V\,d\theta \qquad (9)$$

where U_2 and U_1 are the values of the internal energy at the same volume at temperatures θ_2 and θ_1. This means that if we measure the heat capacity at constant volume over a temperature range we can then find the change in internal energy of the system within that range. This result is quite general.

The internal energy, U, depends only on the thermodynamic state of the body; this, in turn, depends only on, say, the temperature, θ, and volume, V, if the mass is fixed. Consequently, we can write for any arbitrary change in U brought about by changes $d\theta$ and dV:

$$dU = \left(\frac{\partial U}{\partial \theta}\right)_V d\theta + \left(\frac{\partial U}{\partial V}\right)_\theta dV \qquad (10)$$

This equation simply exploits the fact that any change in U depends only on the initial and final states and not on the path; thus we can first change θ with V constant and then change V with θ constant and this will describe any arbitrary infinitesimal change in U we please.

However, we have already seen that $(\partial U/\partial \theta)_V = C_V$ so that we can rewrite equation (10) as

$$dU = C_V d\theta + \left(\frac{\partial U}{\partial V}\right)_\theta dV \tag{11}$$

Experiments on gases at very low pressures show that in them the internal energy at a given temperature does not depend on the volume; this is known as Joule's law. If we regard an ideal gas as the limit of an actual gas at very low pressures, we can conclude that *for an ideal gas* $(\partial U/\partial V)_\theta = 0$ and so, for any change in such a gas,

$$dU = C_V d\theta \tag{12}$$

This is NOT true for most systems; in solids, for example, the internal energy depends strongly on the volume and the second term in equation (11) cannot be neglected.

Exercises

Q1   A gas, not an ideal gas, is contained in a thermally insulated cylinder. It is quickly compressed so that its temperature rises sharply. Has there been a transfer of heat to the gas? Has work been done?

Q2   A gas, not an ideal gas, is contained in a rigid thermally insulated container. It is then allowed to expand into a similar container initially evacuated. What is the change in the internal energy of the gas?

Q3 $10^{-3} \, \mathrm{m}^3$ of lead is compressed reversibly and adiabatically from 1 to 1000 atmospheres pressure. What is the change in the internal energy of the lead? The adiabatic compressibility of lead is assumed independent of pressure with the value of $2.2 \times 10^{-6} \, \mathrm{atm}^{-1}$. Take $1 \, \mathrm{atm} = 10^5$ Pa. (Cf. Exercise Q3 of Chapter 2.)

Q4   A system, whose equation of state depends only on pressure p, volume V and temperature θ, is taken quasi-statically from state A to state B in the figure along the path ACB at the pressures indicated. In this process 50 J of heat enter the system and 20 J of work are done *by* the system.

(a) How much heat enters the system along the path ADB?
(b) If the system goes from B to A by the curved path indicated schematically on the figure, the work done *on* the system is 25 J. How much heat enters or leaves the system?

24

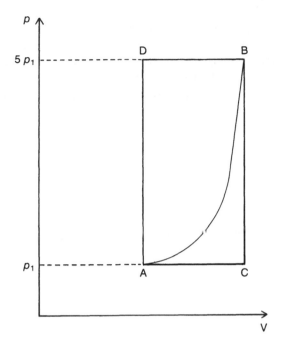

Figure Q4

(c) If the internal energy at A is denoted by U_A etc., suppose that $U_D - U_A = 25$ J. What then is the heat transfer involved in the processes AD and DB?

Q5 1 kg of water is slowly frozen at constant temperature and a pressure of 100 atm. The latent heat of melting of ice under these conditions is 3.36×10^5 J kg^{-1}. How much work is done and what is the change in internal energy of the system? The density of ice relative to that of water is 0.9.

Q6 Show that in an ideal gas:

$$C_p = C_V + R$$

where C_p, the heat capacity per mole at constant pressure, is defined as $C_p = (\text{d}q/\text{d}\theta)_{p=\text{const}}$ and R is the gas constant per mole, as in the equation of state of the gas: $pV = R\theta$.

(Hint: use equations (12) and (7) of the text.)

Q7 Show that, when an ideal gas undergoes an adiabatic quasi-static chanage, $pV^\gamma = $ constant, where $\gamma = C_p/C_V$.

(Hint: use equation (7) of the text, with $dq = 0$ for an adiabatic process, and equation (12) for dU in an ideal gas. Then use the differential form of $pV = R\theta$, i.e. $p\,dV + V\,dp = R\,d\theta$ to eliminate $d\theta$. You will also need the result derived in the previous question Q6.)

Show also that the result can be re-expressed as $\theta V^{\gamma-1} = \text{constant}$.

4

The second law of
thermodynamics

In our discussion of the first law of thermodynamics we have seen how the internal energy function is defined and from that how 'heat' is defined and measured. We have also noted that the internal energy can be identified with the total mechanical energy (kinetic and potential) of the atoms in the substance.

The second law of thermodynamics introduces a further state variable, the entropy. In addition, it makes possible the definition of an *absolute* scale of temperature, that is, one which is independent of the properties of any substance or class of substances. (The *size* of the degree is, however, still defined in terms of the properties of a particular substance, usually water. See page 36.) The second law starts from the following fact of common observation: when a hot body and a cold body are brought into thermal contact, the hot body cools and the cold body gets warm and never the other way round. We may say, in brief, that heat always flows spontaneously from hot to cold. The second law of thermodynamics attempts to express this fact in such a way that it can be used to deduce certain important properties common to all substances.

By the beginning of the nineteenth century, steam engines were widely used in industry and a great deal of work was being done to improve their performance. It was this interest which led to the formulation and understanding of the second law of thermodynamics. Carnot (1796–1832) was the first person to grasp the essentials of the problem and we shall examine his approach to it in some detail.

Before doing so, however, I would like to comment briefly on the relevance of heat engines to the wider topic of thermodynamics. Thermodynamics serves two quite different and at first sight unrelated purposes. On the one hand, it deals with the engineering problem of making heat engines (and refrigerators) work efficiently. On the other hand, it deals with the conditions of *equilibrium* in thermodynamic systems. How is it that these two apparently different subjects are so closely linked? The reason is this (I quote from Carnot): 'Wherever there is a difference of temperature, work can be produced from it.' More generally, whenever two systems are not in thermodynamic equilibrium work can be produced. The expression, 'not in thermodynamic equilibrium', can include a difference in temperature, a difference in pressure, a difference in chemical potential, and so on. So we see that, just as problems of stability and equilibrium in mechanics are related to the ability of a system to do work, so in thermodynamics the possibility of work and the absence of equilibrium are two aspects of the same situation. So when we study the behaviour of idealised heat engines, remember that the results will also be important in the study of the general thermodynamic properties of systems in equilibrium.

With this in mind let us now look at Carnot's work on the behaviour of heat engines.

The behaviour of idealised heat engines

In 1824 Carnot published his famous memoir entitled *Réflexions sur la puissance motrice du feu et sur les machines propres à développer cette puissance*. In this truly remarkable work he was concerned with the general question of how to produce mechanical work from sources of heat. He begins as follows:[8]

> Everybody knows that heat can cause movement, that it possesses great motive power: steam engines so common today are a vivid and familiar proof of it. . . . The study of these engines is of the greatest interest, their importance is enormous, and their use increases every day. They seem destined to produce a great revolution in the civilised world. . . .
>
> Despite studies of all kinds devoted to steam engines, and in spite of the satisfactory state they have reached today, the theory of them has advanced very little and the attempts to improve them are still directed almost by chance.
>
> The question has often been raised whether the motive power of heat is limited or if it is boundless; whether possible improvements in steam engines have an assignable limit, a limit that, in the nature of things, cannot be exceeded by any means whatever, or if on the contrary these improvements can be extended indefinitely.

Carnot adds a footnote here to explain what he means by 'motive power'. He means essentially what we would call 'work'. In the next paragraph he continues:

> To see in its full generality the principle of the production of work by heat we must think of it independently of any particular agent; we must

establish arguments applicable not only to steam engines but to any imaginable heat engine whatever the working substance and whatever its manner of working.

After some further general remarks, Carnot now comes to a most important point. Let me first give a literal translation of what he says and then comment on it.

> The production of motion in the steam engine is always accompanied by a circumstance upon which we must fix attention. This circumstance is the re-establishment of equilibrium in the caloric, that is to say, its passage from a body where the temperature is more or less elevated to another where it is lower.

The expression, 're-establishment of equilibrium in the caloric', can be translated quite accurately by the phrase, 'return to thermal equilibrium'. So the point he is making here is that the production of work by a heat engine requires a temperature difference in the first place but that when the engine operates it will reduce the temperature difference and tend to bring about thermal equilibrium.

Again, later on, Carnot reiterates that the heat engine requires both a hot body and a cold body to make it operate. In addition, and this is most important, 'Wherever there exists a difference of temperature it is possible to produce work from it.'

Carnot then continues his argument as follows:

> Since any return to thermal equilibrium can be used to produce work, any return to equilibrium which takes place without the production of this work must be considered as a real loss.

The point here is that, given a temperature difference, this may either be used to produce work or it may be wastefully dissipated simply by a flow of heat from a high temperature to a low temperature without any work being produced. As we all know, this spontaneous flow of heat from a hot body to a colder one will always take place if it can. This means that a situation which might produce work can be irretrievably lost by natural heat flow.

Let me sum up the situation at this stage. (i) A heat engine requires a temperature *difference* in order to operate. (ii) When the engine operates it takes in heat at the high temperature and gives out some heat at the low temperature so that it tends to reduce the temperature difference, i.e. to restore thermal equilibrium. (iii) Any temperature difference can, in principle, be used to produce work. (iv) Temperature differences tend to disappear spontaneously by heat conduction without producing useful work.

These ideas form the basis of Carnot's thinking (although perhaps with a slightly different emphasis since at the time he wrote he was making use of the caloric theory, even though he had serious doubts about its validity).

Entropy and its physical meaning

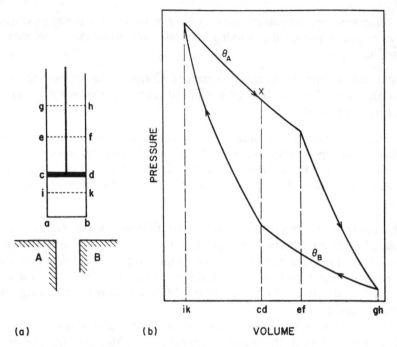

Figure 1 (a) Copy of Carnot's original diagram. (b) The Carnot cycle on the p–V diagram of a gas. θ_A is the temperature of the heat source and θ_B that of the heat sink. The letters on the V axis correspond to the positions of the piston in Figure 1(a). X is the starting point.

From these ideas it is clear that an efficient heat engine must be so designed that there are no wasteful heat flows during its operation. When heat has to be transferred from, say, the furnace to the working substance in the engine, this must be done while both bodies are very nearly at the same temperature. Moreover, there must be no significant temperature gradients inside the working substance at any time during its operation since these too would reduce the efficiency of the engine.

Carnot devised a cycle of operations for a heat engine that would achieve these purposes. This cycle (now called the Carnot cycle) has been described and considered so often since Carnot's time that it is of interest to see how he himself described it in the first place. It is a very lucid description but it may perhaps be helped by showing the cycle of operations on the p–V diagram of a gas. Figure 1 shows such a diagram (Figure 1(b)) together with the original Figure 1 of Carnot's paper (Figure 1(a)). The lettering on the V axis in Figure 1(b) has been chosen to correspond with that of the positions of the piston in Figure 1(a). The cycle shown in the figure consists of two isothermal and two adiabatic processes in the sequence described by Carnot. Notice, however, that Carnot starts his description of the process at the point X, i.e. part way through an isothermal change. Now let us follow Carnot:

30

... Let us imagine a gas, atmospheric air for example, enclosed in a cylindrical vessel *abcd*, Figure 1 [our Figure 1(a)], furnished with a movable diaphragm or piston *cd*; in addition let there be two bodies, *AB*, each held at a constant temperature, that of *A* being higher than that of *B*; let us now imagine the following series of operations:

1   Contact of the body *A* with the air enclosed in the space *abcd*, or with the wall of this space, a wall which we will suppose to conduct heat easily. Through this contact the air assumes the same temperature as the body *A*; *cd* is the present position of the piston.

2   The piston rises gradually and takes the position *ef*. Contact is still maintained between the body *A* and the air, which is thus kept at a constant temperature during the expansion. The body *A* furnishes the heat necessary to keep the temperature constant.

3   The body *A* is taken away and the air is no longer in contact with any body capable of giving it heat; the piston however continues to move and passes from the position *ef* to the position *gh*. The air expands without receiving heat, and its temperature falls. Let us imagine that it falls in this way until it becomes equal to that of the body *B*: at this stage the piston stops and occupies the position *gh*.

4   The air is put into contact with body *B*; it is compressed by the return of the piston as it is moved from the position *gh* to the position *cd*. The air however remains at a constant temperature because of its contact with the body *B* to which it gives up its heat.

5   The body *B* is removed and we continue to compress the air which, being now isolated, rises in temperature. The compression is continued until the air has acquired the temperature of the body *A*. During this time the piston passes from the position *cd* to the position *ik*.

6   The air is put back into contact with body *A*; the piston returns from the position *ik* to the position *ef*; the temperature stays constant.

7   The operation described under No. 3 is repeated, then successively the operations 4, 5, 6, 3, 4, 5, 6, 3, 4, 5 and so on. ...

Carnot notes two points about this cycle. One is that the working substance (the air) produces a net amount of work in each cycle. We can see this readily from the *p–V* diagram. The work done *by* the air is $\int p\,dV$ which is thus the area of the cycle on the *p–V* diagram. If the representative point of the working substance traces out the cycle in the direction indicated by the arrows, this area is positive and a net amount of work is produced in each cycle.

The second point is that the work is produced in the most advantageous manner possible. The two adiabatic processes (3) and (5) change the temperature of the air in the engine without adding or taking away any heat. In this way the air is always brought to the temperature of the heat source before being put into contact with it, and likewise the heat sink. This prevents any wasteful flow of heat between bodies at widely different temperatures.

31

Entropy and its physical meaning

From our point of view we may say that all the processes described in the cycle are quasi-static; the gas is always effectively in equilibrium. For this reason all the changes are reversible which is important for the subsequent argument.

Carnot continues:

> All the operations described above can be carried out in the opposite direction and in an inverse order. Let us imagine that after the sixth operation, that is to say the piston having arrived at the position *ef*, one makes it return to the position *ik* and that at the same time one keeps the air in contact with body *A*: the heat furnished by the body during the sixth operation will return to its source, that is to say to body *A*, and things will be in the state they were at the end of the fifth operation. If now we take away the body *A* and make the piston move from *ef* to *cd*, the temperature of the air will decrease by as many degrees as it rose during the fifth operation and will become equal to that of body *B*. We can obviously continue a series of operations inverse to those that we first described: it is sufficient to reproduce the same conditions and to carry out for each operation an expansion instead of a compression and *vice versa*.
>
> The result of the first operations was the production of a certain quantity of work and the transport of heat from body *A* to body *B*; the result of the inverse operations is the consumption of the work produced and the return of the heat from body *B* to body *A*: so that these two series of operations in a way cancel or neutralise each other.

Carnot then shows that no heat engine can be more efficient than the reversible engine he has described. He remarks:

> We have chosen atmospheric air as an instrument for developing work from heat; but it is obvious that the reasoning would have been the same for any other gaseous substance and even for any other body whose temperature can be changed by successive contractions and expansions which includes all substances in nature or at least all those which are appropriate for obtaining work from heat. Thus we are led to establish the following general proposition:
>
> *The motive power of heat is independent of the agents used to produce it; its amount is fixed uniquely by the temperatures of the bodies between which the transport of heat is ultimately made.*

We shall not follow Carnot's own argument to establish this proposition because it rests on the caloric theory. To make the argument consistent with the mechanical theory of heat, we must change it somewhat and also bring in a new principle: the second law of thermodynamics.

As we have seen, the Carnot cycle consists of two adiabatic and two isothermal processes (at temperatures θ_A and θ_B) so carried out that at all stages the changes are reversible.

Suppose now that during one cycle the quantity of heat taken in by the working substance at the high temperature is Q_1, and that given out at the low temperature is Q_2. Since the working substance returns exactly to its initial state at the end of one cycle, $\Delta U = 0$. If then the work done *by* the engine in one cycle is W,

$$W = Q_1 - Q_2 \tag{13}$$

by virtue of the first law of thermodynamics. The ratio of the work done per cycle to the heat taken in at the high temperature, i.e. W/Q_1, is called the thermodynamic efficiency of the engine. We now wish to show that all reversible engines working between the same two temperatures have the same efficiency and further that no engine working between these temperatures can be more efficient.

In order to proceed further, we require to state in a suitable form the axiom or postulate that heat always flows spontaneously from a high temperature to a lower one. Clausius (1822–1888), in his statement of the second law of thermodynamics, used the following form: 'It is impossible for a self-acting (cyclic) machine, unaided by any external agency, to convey heat from one body at a given temperature to another at a higher temperature.'[9] The form of the second law of thermodynamics due to Kelvin (William Thomson, 1824–1907) is as follows:[10] 'It is impossible, by means of inanimate material agency, to derive [continuous] mechanical effect from any portion of matter by cooling it below the temperature of the coldest of the surrounding objects.' These two statements can be shown to be completely equivalent but the first is perhaps preferable as being more obviously a restatement of the observation that heat flows spontaneously only from hot to cold.

We wish now to prove that no engine using a given source and sink of heat can be more efficient than a reversible engine (a Carnot engine) working between the same two temperatures. Let A and B be two engines of which A is the Carnot engine and B is the engine which, by hypothesis, has a greater thermodynamic efficiency than A. Since A is perfectly reversible, we may suppose that it is driven backwards absorbing completely the work, W, generated in each cycle by B. A is now no longer a heat engine but a refrigerator or heat pump whose function is to absorb heat at the low temperature and discharge a greater quantity at the higher temperature. If B absorbs heat Q_b at the high temperature and A restores heat Q_a, then the thermodynamic efficiency η_a of A is W/Q_a, and that of $B(\eta_b)$ is W/Q_b. Now by hypothesis η_b is greater than η_a since we supposed B to be more efficient than A. This implies that Q_a is greater than Q_b. That is to say that more heat is delivered by A at the high temperature than is absorbed by B. This is no violation of the first law of thermodynamics since A absorbs at the low temperature correspondingly more heat than is discharged by B. It is, however, a violation of the second law of thermodynamics since the two machines together constitute 'a self-acting cyclic machine' which 'unaided by any external agency . . . conveys heat from one body at a low temperature to another at a higher'. We conclude, therefore, that our original hypothesis is wrong, i.e. that no heat engine with an efficiency greater than a Carnot engine can exist. From a similar argument it follows that all Carnot engines working between the same two temperatures have the same efficiency. If this were not so it would be possible, by suitably coupling a pair of engines whose efficiencies were different, to violate the second law of thermodynamics.

We have established, therefore, that the thermodynamic efficiency of a Carnot engine depends only on the temperatures between which it works and not on the working substance. This is essentially the proposition which Carnot set out to prove; it is frequently called Carnot's theorem.

Although we have not followed exactly Carnot's original argument, we have followed almost all its essential features and in particular the idea of coupling the reversible engine, working backwards, to the heat engine whose efficiency is under study.

How did Carnot come upon such beautiful, general and powerful arguments? Part of the answer lies in the influence and achievements of his father, Lazare Carnot, who was not only one of Napoleon's most successful generals, but was also a very capable mathematician. One of his mathematical works, *Fundamental Principles of Equilibrium and Movement*, was concerned with the efficiency of machines such as pulleys and levers. In the preface, he writes of mechanical engines in general (I translate): '. . . to make these [machines] produce the greatest effect possible, it is essential that there are no sudden jolts, that is to say that the movement must always change by imperceptible degrees.' In the text he amplifies this and in discussing how to make an efficient mechanical engine writes: '. . . this principle demands that one avoids . . . any collision or sudden change whatever that is not essential to the nature of the machine, since every time there is a collision there is a loss of kinetic energy, and in consequence a part of the active movement is uselessly absorbed.' If you compare this with Sadi Carnot's words on heat engines (see p. 36), you see, I think, how closely the grain of Sadi's thinking matches that of his father. Nonetheless his son's work on heat engines carries Lazare's ideas into quite new realms. Sadi Carnot's mode of thought is further revealed in the text of his treatise. At one point he says (this is a fairly literal translation and not very elegant):

> According to our present ideas, one can, with sufficient accuracy, compare the motive power of heat with that of water descending: both have a maximum that cannot be exceeded . . . no matter what machine is used to absorb the action of the water or . . . what substance is used to absorb the action of the heat. The motive power of the fall of water depends on its height and the quantity of liquid; the motive power of heat likewise depends on the quantity of heat used and what we might call, and what indeed we shall call, *the height of its fall*, that is to say, the difference in temperature of the bodies between which the exchange of heat takes place.

Carnot's analogy turned out not to be exact but it provided an astonishingly clear insight into the working of heat engines and from that to a profound understanding of thermodynamics.

We now proceed to do two things: (i) to define the absolute temperature scale; (ii) to define a new state variable, the entropy.

The absolute scale of temperature

We have already shown that the efficiency of a Carnot engine whose working temperatures are t_1 and t_2 is a function only of these temperatures, i.e.

$$\eta = \frac{Q_1 - Q_2}{Q_1} = 1 - \frac{Q_2}{Q_1} = \Phi(t_1, t_2) \tag{14}$$

t_1 and t_2 are *not* now measured on the gas scale of temperature; they are at this point quite arbitrary. Therefore Q_2/Q_1 is itself simply a function of t_1 and t_2, i.e.

$$\frac{Q_2}{Q_1} = F(t_2, t_1) \tag{15}$$

Moreover, it is not difficult to show that the function F must separate in such a way that $Q_2/Q_1 = f(t_2)/f(t_1)$. This then provides us with a general basis for an absolute scale of temperature. Note, however, that we may obviously choose the function f as we please.

Thomson (later Lord Kelvin) recognised even before the general acceptance of the present theory of heat, that the fact expressed in Carnot's theorem could provide the basis for an absolute scale of temperature. He first proposed such a scale in 1848 defining it in the following way:[11]

> The characteristic property of the scale which I now propose is, that all degrees have the same value; that is, that a unit of heat descending from a body *A* at the temperature $T°$ of this scale, to a body *B* at the temperature $(T-1)°$, would give out the same mechanical effect, whatever be the number *T*. This may justly be termed an absolute scale, since its characteristic is quite independent of the physical properties of any specific substance.

This turns out to be equivalent to choosing the function $f(t)$ as the exponential function so that in a Carnot cycle

$$\frac{Q_1}{Q_2} = \frac{e^{T_1^\circ}}{e^{T_2^\circ}} \tag{16}$$

where T_1° and T_2° are the temperatures of the sink and source on this first Thomson scale.

In 1854 Thomson defined a second absolute scale, chosen this time to agree with the air thermometer scale, which is effectively the gas scale of temperature.[11] In this second scale, which is now universally used and called the Kelvin temperature scale, the temperatures T_1 and T_2 of the source and sink of the Carnot engine are so defined that:

$$\frac{Q_1}{Q_2} = \frac{T_1}{T_2} \tag{17}$$

This equation defines the *ratio* of two absolute temperatures. It does not define the size of the degree. To do this we must give a numerical value to the temperature of some convenient fixed point. For this purpose the triple point of water has been chosen as the most convenient and its temperature is taken to have the value 273.16 K *by definition*. (This choice was made and recommended by the Tenth General Conference on Weights and Measures in 1954.) The reason for choosing this particular number is that the size of the degree then remains essentially the same as the Centigrade degree used previously in scientific work. Note, however, that whereas to use the Centigrade degree requires measurements at two fixed points (the ice point and the normal boiling point of water) to use the degree defined in this new way requires measurements at one fixed point only.

From now on we shall always use this absolute temperature scale. It is usually denoted by K. By taking a perfect gas as the working substance of a Carnot engine it is not difficult to show that the gas scale is then identical with the absolute scale. Over a wide range of temperatures the gas scale can thus be used as a means of determining the absolute temperature. At very low temperatures, however, gases effectively cease to exist and then the gas scale can no longer be used. In this region, as we shall see, we have to go back to the definition of the absolute scale in order to measure the temperature.

The Carnot engine as a refrigerator

A Carnot engine, being perfectly reversible, is both the most efficient heat engine and also, when operating in reverse, the most efficient refrigerator or heat pump. It can thus be used as a standard with which to compare actual refrigerating machines.

However, we must beware of thinking that the idealised efficiency defined above has any close relationship to 'efficiency' as an engineering concept. To the engineer many other aspects of the engine, whether as a heat engine or refrigerator, that are neglected in the idealisation are of importance, probably of overriding importance: speed of operation, cost, environmental effects and so on.

Exercises

Q1 An ideal gas with $\gamma = 1.5$ is used as the working substance of a Carnot engine. The temperature of the source is 600 K and that of the sink is 300 K.

The volume of the gas changes from 4 to 1 litre at the low temperature and the pressure at the volume of 4 litres is 1 atmosphere. Calculate the volumes and pressures at the extreme points of the high temperature expansion and also the heat absorbed there. How much heat is given out at the low temperature and what is the thermodynamic efficiency of the engine?

Q2   A house is heated by means of a reversible heat engine acting as a heat pump. The outside temperature is 270 K and the inside temperature is 300 K. If when the house is heated by conventional electric space heaters 10 kW of power is used, how much power is consumed by the ideal heat pump to maintain the same temperature?

Q3   Two identical bodies of constant heat capacity are initially at temperatures T_1 and T_2. They are used as the sink and source of a Carnot engine and no other source of heat is available. Show that the final common temperature when all possible work has been extracted from the system is $T_F = (T_1 T_2)^{1/2}$. What is the maximum work obtainable?

Q4   The volume thermal expansion coefficient of helium gas at the ice point and at low pressures is 0.003 66 per °C. The temperatures were measured on the perfect gas scale. What is the value of the ice point on the absolute temperature scale?

5

Entropy: how it is measured and how it is used

The definition of entropy

So far Q_1 and Q_2 have been considered as magnitudes only. However, if we adopt a sign convention by which Q is positive when heat is absorbed by the working substance and negative when heat is given up, we may rewrite equation (17) in the form

$$Q_1/T_1 + Q_2/T_2 = 0 \tag{18}$$

Figure 2 represents in a p–V diagram the succession of states through which a substance is taken in an arbitrary reversible cycle: superimposed on it are drawn a succession of adiabatic and isothermal paths. The heavy line in the diagram represents a path consisting only of alternating adiabatic and isothermal steps, which approximate to the original arbitrary cycle and can, in fact, be made as close to this as we wish by making the mesh of adiabatic and isothermal lines still finer. (This is not obvious because the zig-zag path is necessarily longer than the smooth one; the first law of thermodynamics, however, is sufficient to show that the heat absorbed in any segment of the smooth path is the same as that absorbed in the corresponding isothermal step, when this is correctly defined.)

Any appropriate set of two isothermal and two adiabatic lines such as, for example, 1, 2, 3, 4 in the figure constitute a Carnot cycle; consequently if q_1 is the heat absorbed at the high temperature and q_2 that emitted at the low temperature in this cycle, $q_1/T_1 + q_2/T_2 = 0$. In a similar way all the other isothermal parts may be paired off, and since no heat is absorbed or emitted in the adiabatic parts of the cycle, we have by summing over all the isothermal steps

$$\sum \frac{q_i}{T_i} = 0 \tag{19}$$

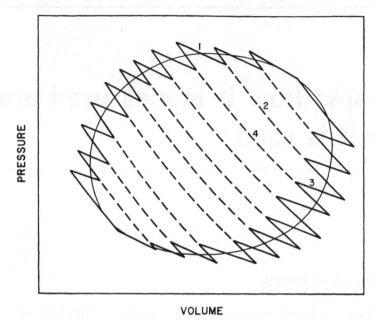

Figure 2 An arbitrary reversible cycle represented in a *p–V* diagram. The zig-zag line is an equivalent path consisting of adiabatic and isothermal steps. The four lines 1, 2, 3 and 4 constitute a Carnot cycle.

This is true for any complete reversible cycle. If now we proceed to the limit of infinitesimally small divisions of the path into adiabatic and isothermal steps, we may write

$$\oint \frac{\mathrm{d}q}{T} = 0 \tag{20}$$

where the integration is carried over the complete reversible cycle. This is Clausius' theorem; it was enunciated by him in 1854.[12]

If we now define a quantity S, the entropy, such that $\mathrm{d}S = \mathrm{d}q_{rev}/T$ ($\mathrm{d}q_{rev}$ means an infinitesimally small amount of heat entering or leaving the system reversibly) we see that $\oint \mathrm{d}S = 0$. In words: a small change in entropy $\mathrm{d}S$ is defined as equal to the small quantity of heat entering or leaving the body reversibly divided by the absolute temperature of the body. If heat enters the body its entropy increases and $\mathrm{d}S$ is positive; if heat leaves the body its entropy decreases and $\mathrm{d}S$ is negative.

In any closed reversible cycle

$$\oint \mathrm{d}S = 0 \tag{21}$$

This implies that the entropy difference between two states of a system depend only on those states and *not* on the path between them. To see this, let A be the initial and end point of a closed, reversible cycle (specified in the simplest case by a point on the p–V diagram) and let B be some arbitrary point on the cycle. We can then re-express equation (21) above as

$$\oint dS = \underbrace{\int_A^B dS}_{(1)} + \underbrace{\oint_B^A dS}_{(2)} = 0$$

where the first integral is by route 1 and the second by the return route 2. Thus

$$\underbrace{\int_A^B dS}_{(1)} = -\underbrace{\int_B^A dS}_{(2)} = \underbrace{\int_A^B dS}_{(2)}$$

Since this is true of any arbitrary cycle through A and B it follows that $\int_A^B dS$ is independent of the path between A and B provided only that it is reversible. Thus the entropy change of the substance between states A and B is independent of the path. This is a very remarkable result. Although the heat entering the substance depends on the path followed by the substance between A and B, the entropy change does not.

This result is so important and, at first sight, so surprising that it may be helpful to illustrate it with a specific example. In this, we shall see directly that while $\int đq_{rev}$ is indeterminate until the path is known, $\int đq_{rev}/T$ is determined entirely by the beginning and end points of the process. The simplest example which demonstrates the change is provided by the behaviour of a perfect gas. According to the first law of thermodynamics:

$$dU = -p\,dV + đq$$

in any quasi-static change. Furthermore, as we saw earlier, the perfect gas has the simple property that in any change of state $dU = C_V dT$. The infinitesimal amount of heat which enters the gas in an infinitesimal quasi-static change of any kind is therefore given by:

$$đq = C_V dT + p\,dV \tag{22}$$

To find the heat entering the gas during any arbitrary finite change we need to integrate this equation giving

$$Q = \int_{T_1}^{T_2} C_V dT + \int_{V_1}^{V_2} p\,dV$$

$$Q = C_V(T_2 - T_1) + \int_{V_1}^{V_2} p\,dV$$

assuming for simplicity that C_V is independent of T, although this is not a

41

necessary assumption. Unfortunately the second term, $\int_{V1}^{V_2} p\,dV$, cannot be evaluated until we know how p varies as the change proceeds. Obviously we do not know this in general and as soon as we do know it, for instance by specifying that p is to remain constant, we are limited to a particular class of changes, in this example isobaric changes. To evaluate the integral we must specify the path; for every path there is in general a different value of the integral. Consequently the same thing is true of the value of Q.

Now contrast this with what happens when we divide equation (22) by T, the absolute temperature. Then

$$dS = \frac{\text{d}q}{T} = \frac{C_V dT}{T} + \frac{p\,dV}{T}$$

so that

$$\Delta S = \int_{T_1}^{T_2} \frac{C_V dT}{T} + \int_{V_1}^{V_2} \frac{p\,dV}{T} \tag{23}$$

Both integrals can now be evaluated without knowing anything about the path. The equation of state $pV = RT$ (for one mole of gas) enables us to transform the second integral of the right-hand side into

$$\int_{V_1}^{V_2} \frac{R\,dV}{V} = R \ln (V_2/V_1)$$

so that

$$\Delta S = C_V \ln (T_2/T_1) + R \ln (V_2/V_1) \tag{23A}$$

which depends *only* on the initial and final states of the gas.

Note that this result which we have just illustrated in a very simple case – namely, that the entropy change depends only on the initial and final states of the system – is quite general. Notice also the close connection between the definition of the absolute temperature and the definition of entropy. Suppose, for example, we had used Thomson's first absolute temperature scale instead of the second. The first scale was defined so that in a Carnot cycle:

$$\frac{Q_1}{Q_2} = \frac{e^{T_1^{\circ}}}{e^{T_2^{\circ}}}$$

where Q_1 is the heat absorbed at the temperature T_1° and Q_2 that given out at T_2°. Then the definition of the small entropy change dS associated with heat $\text{d}q$ entering a system reversibly would have had to be $dS = \text{d}q/e^{T^{\circ}}$. In short, the definitions of absolute temperature and of entropy cannot be separated.

The concept of entropy was introduced by Clausius in 1854; the name 'entropy' was introduced by him considerably later, in a paper of 1865.[13]

The measurement of entropy

At present we shall concentrate on the measurement of *changes* in entropy although later we shall see that there is a natural zero for entropy which has a special significance.

The specific heat capacity of a substance is defined as 'the heat required to raise unit mass of the substance by one unit of temperature'. More exactly the specific heat capacity C is the value of the ratio $q/\delta T$ in the limit that δT tends to zero. Here q is the heat required to raise the temperature of unit mass of the substance by δT. Thus in a reversible change $đq_{rev} = C \, dT$, and the associated entropy change is $dS = đq_{rev}/T = C \, dT/T$. If we invert this relation we get $C = T \, dS/dT$ or, to take two particular examples:

$$C_V = T\left(\frac{\partial S}{\partial T}\right)_V$$

$$C_p = T\left(\frac{\partial S}{\partial T}\right)_p$$

You may object here that in most specific heat measurements the process is in fact irreversible, that the energy is usually supplied by passing a current through a resistance which is an irreversible process. Moreover, heat capacities are nearly always measured with increasing temperature and, in the common methods, measurements with falling temperature are impossible. What, then, does q_{rev} mean? A careful examination of this question will show that the criterion of reversibility (as far as the temperature change is concerned) only requires that the temperature *interval* be kept small enough, i.e. that $\delta T/T$ in the expression for the entropy change be smaller than the required accuracy of the result. This ensures that the addition of heat satisfies the requirements of reversibility; besides this, however, you have to ensure that no other irreversible processes take place during the temperature change. In glasses or ferromagnetics, for example, where thermal hysteresis may be present this latter condition may be violated and then you cannot deduce the entropy change from the apparent specific heat. When we come to the third law of thermodynamics, we will discuss the question of glasses again.

If we know the heat capacity of a system over a temperature range, we can then find the corresponding entropy change. This is given by:

$$S_2 - S_1 = \int_{T_1}^{T_2} C \, dT/T \tag{24}$$

(If C is constant, this can be integrated at once and we see that under these circumstances S varies as $\ln T$.) If C is measured at constant volume, S_2 and S_1 refer to that volume. If C is measured at constant pressure, S_2 and S_1 then refer to that pressure. In many cases the measurement of an entropy change thus involves simply the measurement of the heat capacity of the substance.

We now know how entropy is defined and, in a general way, how it can be measured. We do not, however, at this stage have any real idea of what entropy *is*. When later we come to interpret entropy on the atomic scale we shall gain a more intuitive idea of the meaning of entropy and what exactly is the property of a body which corresponds to the entropy as already defined. In the meantime, however, we shall work towards a qualitative grasp of what entropy is by examining some examples of how it behaves under different conditions. To help in this, let me first explain briefly what entropy–temperature diagrams are and how they can be used.

Entropy–temperature diagrams

The thermodynamic state of a system may be indicated on an entropy–temperature diagram in the same way as in a *p–V* diagram. Similarly, quasi-static processes can be represented by lines in this diagram. An adiabatic reversible change is a reversible change in which no heat enters or leaves the system; $đq_{rev}$ is thus zero and dS likewise; so S is constant. This kind of change can therefore be represented on an *S–T* diagram as a straight line at constant S, i.e. parallel to the *T*-axis. Similarly an isothermal reversible process is represented by a straight line parallel to the *S*-axis. A Carnot cycle is thus represented by a rectangle as shown in Figure 3.

Since in a reversible change $đq = T\,dS$, the total amount of heat absorbed in changing the entropy of a system reversibly from S_1 to S_2 is

$$Q = \int_{S_1}^{S_2} T\,dS \tag{25}$$

This is just the area between the line representing the change from S_1 to S_2 and the *S*-axis. For the change *AD* it is the shaded area in Figure 3. This result holds for any reversible path; for example, in the Carnot cycle the heat given out, Q_1, in the isotherm at T_1 is $T_1(S_2 - S_1)$ and the heat absorbed, Q_2, in the isotherm at T_2 is $T_2(S_2 - S_1)$. Thus $Q_1/Q_2 = T_1/T_2$ as, of course, we know from the definition of the absolute temperature.

Now let us turn to some examples of how the entropy of various systems behaves; in what follows we shall frequently make use of entropy–temperature diagrams.

The entropy of a perfect gas

We have already seen that if C_V is assumed constant, the difference between the entropy per mole of a perfect gas at temperature T_2 and volume V_2 and

44

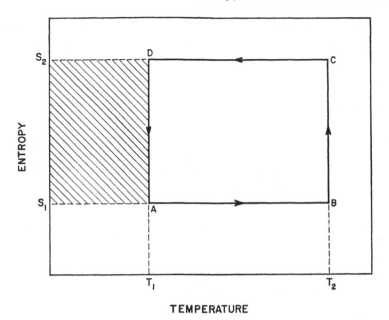

TEMPERATURE

Figure 3 A Carnot cycle represented on an entropy–temperature diagram of the working substance. The heat absorbed by the working substance in going from A to D is represented by the shaded area. In going from D to A this amount of heat is given out.

that of the gas at T_1 and V_1 is:

$$S_2 - S_1 = C_V \ln \frac{T_2}{T_1} + R \ln \frac{V_2}{V_1} \tag{26}$$

We can therefore write for the entropy of the gas at temperature T and volume V:

$$S = C_V \ln T + R \ln V + \text{constant} \tag{27}$$

We thus see that if the volume is constant, S increases as $\ln T$; while if the temperature remains constant, S increases as $\ln V$. It is interesting to compare this behaviour with that of the internal energy, U. For an ideal gas whose heat capacity at constant volume is independent of temperature

$$U = C_V T + \text{constant} \tag{28}$$

This therefore depends linearly on T and is independent of V.

Returning to equation (27), we see that if we compress a gas isothermally and reversibly the entropy decreases. This corresponds to the fact that we have to remove heat from the gas if its temperature is not to rise. We may think of the process as one in which we 'squeeze' the entropy out of the gas; this entropy shows itself as the heat of compression. Likewise a reversible expansion of the gas requires that heat be supplied to the gas if it is not to

45

become cooler in the process. As we shall see later a slow expansion under *adiabatic* conditions is a very effective way of producing low temperatures.

Entropy change and latent heat

In a phase change such as the melting of a solid or the vaporisation of a liquid at a constant pressure, the complete transformation takes place at a constant temperature; during the transformation heat must be supplied (or removed) and this is the latent heat of melting or of vaporisation, L. The entropy change associated with the phase change is therefore

$$\Delta S = \frac{L}{T}$$

where T is the temperature of the phase change. L is in general a function of T so that ΔS varies as the temperature and pressure of the transition are altered. Table 1 gives some values of the entropy change on vaporisation ΔS_V for a number of different elements at their normal boiling points. Table 2 gives the entropy change on melting ΔS_m for a number of elements at the temperatures specified.

Let us now look at the entropy–temperature diagram of a liquid and vapour in equilibrium (Figure 4). The area enclosed by the heavy line is the region in which both liquid and vapour co-exist. Consider the vertical line $A–B$. The point A corresponds to the entropy of the vapour in equilibrium with the liquid at this temperature while the point B corresponds to the entropy of the liquid in equilibrium with the vapour. The distance between A and B is just $\Delta S = L/T$, where L is the latent heat of vaporisation at temperature T. A point C on this line, which divides it in the ratio x to y

Table 1 Entropy of vaporisation of various elements

Substance	T_B (K)	L_V (J mole^{-1})	ΔS_V (J mole^{-1} K^{-1})
Helium (He^4)	4.2	92	22
Hydrogen (normal)	20.4	904	45
Neon	27.4	1 760	65
Nitrogen	77.3	5 610	72
Argon	87.3	6 530	75
Oxygen	90.1	6 820	76
Mercury	630	59 500	95
Sodium	1155	97 500	103
Zinc	1180	116 000	103
Lead	1887	193 000	110

Table 2 Entropy of melting of various elements

Substance	T_m (K)	L_m (J mole^{-1})	ΔS_m (J mole^{-1} K^{-1})
He^3	2	8	4
	10	63	6
He^4	2	9	4.6
	10	67	6.7
Neon	24.6	330	13
Argon	83.9	1 180	14
Krypton	116.0	1 630	14
Xenon	161.3	2 300	14
Potassium	336.4	2 320	6.9
Sodium	371.0	2 610	7.0
Lithium	453	3 070	6.8
Copper	1356	13 000	9.6
Silver	1234	12 500	10.0
Gold	1336	13 400	10.0

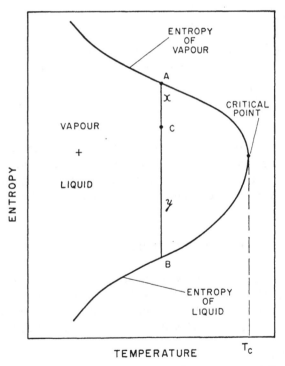

Figure 4 The entropy–temperature diagram of a vapour and liquid in equilibrium. The region to the left of the curved line represents a two-phase region in which the liquid and vapour co-exist.

as shown, indicates that the system represented by this point is a mixture of liquid and vapour in the proportions of x parts of liquid to y parts of vapour. At the temperature marked T_c the entropy difference between the two phases disappears and above this temperature there is no two-phase region at all. This temperature is the critical temperature; at this point, too, the volume difference between the two phases also disappears.

Now let us deduce the relationship between the saturated heat capacities and the entropy change on vaporisation. The saturated heat capacity of a liquid is defined as the heat required to raise the temperature of unit mass by one degree while the liquid remains always in equilibrium with its vapour. The saturated heat capacity of the vapour is defined in an analogous way.

Let S_1 be the entropy of the liquid and S_2 that of the vapour in equilibrium with it at temperature T. If the latent heat of the transition at this temperature is L, we have:

$$S_2 - S_1 = \frac{L}{T} \tag{29}$$

and so:

$$T\frac{dS_2}{dT} - T\frac{dS_1}{dT} = T\frac{d}{dT}\left(\frac{L}{T}\right) \tag{30}$$

Now $T(dS_2/dT)$ is just the heat capacity of the saturated vapour and $T(dS_1/dT)$ is just the heat capacity of the saturated liquid. Therefore:

$$C_{2\,sat} - C_{1\,sat} = \frac{dL}{dT} - \frac{L}{T} \tag{31}$$

This is the Clapeyron equation.

What is important here is to observe that the single quantity, the entropy, and its dependence on temperature, contain all the thermodynamic information about the saturated heat capacity of the liquid, the saturated heat capacity of the vapour, the latent heat and their mutual relationship.

The central equation of thermodynamics

With the introduction of entropy we come to the heart of thermodynamics. We can sum up what we have done so far by bringing together in one very important equation the zeroth, first and second laws of thermodynamics. In an infinitesimal process in a closed system the first law requires that:

$$dU = đw + đq$$

in the usual symbols.

If the change is quasi-static and involves a change in volume dV at pressure, p, $đw = -p\,dV$; moreover, from the second law of thermodynamics, $đq = T dS$, where T is the absolute temperature of the body and dS the entropy change associated with the heat $đq$.

Thus

$$dU = T dS - p dV \tag{32}$$

This important relation (which applies to a simple closed system)[14] involves the zeroth law through the concept of temperature, the first law through the internal energy function and the second law through the entropy and absolute temperature. It is really the central equation of thermodynamics; the applications of thermodynamics consist in examining the consequences of this equation (suitably generalised where necessary) and applying them in a wide variety of circumstances and to a wide variety of systems.[15]

Thermodynamic functions and their appropriate variables

Now let us look at some of the implications of equation (32). This equation tells us that if we know U as a function of S and of $V - U(S, V)$, for short – we know everything we can know about the system from a strictly thermodynamic point of view. We can see this most easily by imagining that we know the function $U(S, V)$ and allowing this to change arbitrarily. Then

$$dU = \left(\frac{\partial U}{\partial S}\right)_V dS + \left(\frac{\partial U}{\partial V}\right)_S dV \tag{33}$$

For any well-behaved function of two variables this is a mathematical identity. (Moreover, for such functions $\partial^2 U/\partial V \partial S = \partial^2 U/\partial S \partial V$, a fact we shall soon use.) Consequently, if we compare this equation with equation (32) and require that the two be identical, we have:

$$T = \left(\frac{\partial U}{\partial S}\right)_V \quad \text{and} \quad p = -\left(\frac{\partial U}{\partial V}\right)_S \tag{34}$$

This means that if we know $U(S, V)$ we can at once find T and p simply by differentiation. In other words, T and p are strictly speaking redundant; they could always be written in the form of equation (34). Of course this would be highly inconvenient in practice because p and T are much easier to measure than, say, S or U. Nevertheless it is important to realise that $U(S, V)$ contains *all* the thermodynamic information about the system that there is. By contrast the equation of state $\theta(V, p)$ does not.

We can derive an additional relationship from equation (32) which is often useful. Since we know that

$$\frac{\partial^2 U}{\partial V \partial S} = \frac{\partial^2 U}{\partial S \partial V}$$

49

we can apply this result to the equation (34). Differentiate the first of these equations with respect to V and the second with respect to S and equate the two. We therefore get:

$$\left(\frac{\partial T}{\partial V}\right)_S = -\left(\frac{\partial p}{\partial S}\right)_V \tag{35}$$

As we shall see below, there are three other analogous identities of this kind.

We have just seen that if we are working with the internal energy function the appropriate variables are S and V. If these are not convenient then a different function can be defined and used. Suppose, for example, we wished to use the variables S and p instead of S and V. For this purpose we introduce a new thermodynamic function of state, called the enthalpy, H, defined by the relationship

$$H = U + pV \tag{36}$$

Since H is a function of U, p and V which are all state variables, H itself must be one.

Thus $dH = dU + p\,dV + V\,dp$ in any arbitrary change. But, from (32)

$$dU + p\,dV = T\,dS$$

so

$$dH = T\,dS + V\,dp \tag{37}$$

Now we can see that the appropriate variables for H are S and p. From equation (37) by repeating the previous procedure we find that

$$\left(\frac{\partial T}{\partial p}\right)_S = \left(\frac{\partial V}{\partial S}\right)_p \tag{38}$$

In a similar way, the Helmholtz free energy, F, is defined by the relation:

$$F = U - TS \tag{39}$$

which leads to

$$dF = -S\,dT - p\,dV \tag{39A}$$

from which $S = -(\partial F/\partial T)_V$ (a result we shall find useful later) and

$$\left(\frac{\partial S}{\partial V}\right)_T = \left(\frac{\partial p}{\partial T}\right)_V \tag{40}$$

The appropriate variables for F are T and V.

Finally, the Gibbs free energy, G, completes the scheme. It is defined as

$$G = U - TS + pV \tag{41}$$

Table 3 Maxwell's thermodynamic relations

Quantity	Symbol and appropriate variables	Definition	Differential relationship	Corresponding Maxwell relation
Internal energy	$U(S, V)$		$dU = T\,dS - p\,dV$	$\left(\dfrac{\partial T}{\partial V}\right)_S = -\left(\dfrac{\partial p}{\partial S}\right)_V$
Enthalpy	$H(S, p)$	$H = U + pV$	$dH = T\,dS + V\,dp$	$\left(\dfrac{\partial T}{\partial p}\right)_S = \left(\dfrac{\partial V}{\partial S}\right)_p$
Helmholtz free energy	$F(T, V)$	$F = U - TS$	$dF = -S\,dT - p\,dV$	$\left(\dfrac{\partial S}{\partial V}\right)_T = \left(\dfrac{\partial p}{\partial T}\right)_V$
Gibbs free energy	$G(T, p)$	$G = U - TS + pV$	$dG = -S\,dT + V\,dp$	$\left(\dfrac{\partial S}{\partial p}\right)_T = -\left(\dfrac{\partial V}{\partial T}\right)_p$

Then

$$dG = -S\,dT + V\,dp$$

so that

$$\left(\frac{\partial S}{\partial p}\right)_T = -\left(\frac{\partial V}{\partial T}\right)_p \tag{42}$$

The appropriate variables for G are T and p; because these are often very convenient variables to work with, the Gibbs free energy is an important thermodynamic parameter.

The equations (35), (38), (40) and (42) (called Maxwell's thermodynamic relations) are all essentially equivalent; you can go from any one of them to any of the others by mathematical manipulation only. Table 3 summarises these Maxwell relations together with the definitions of the various quantities which were introduced in deducing them.

To show how we can apply the results which we have just obtained we shall consider two examples which will be useful to us later on. One is a deduction of the Clausius–Clapeyron equation relating to phase equilibrium; the second is a thermodynamic treatment of the Joule–Thomson effect.

The Clausius–Clapeyron equation

In phase changes such as melting and evaporation, the transition from one phase to the other takes place at a constant temperature if the pressure is kept constant. If we apply our fundamental equation

$$dU = T\,dS - p\,dV$$

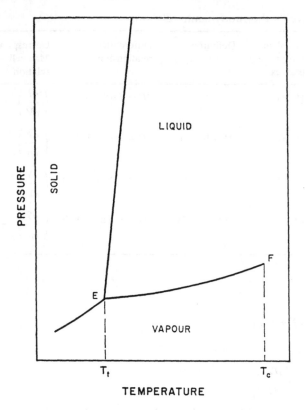

Figure 5 A typical phase diagram for a pure substance. *E* is the triple point at which solid, liquid and vapour can all be in equilibrium together; T_t is the triple point temperature. *F* is the critical point; T_c is the critical temperature.

to the process we can at once integrate this, because p and T do not change. If U_1, S_1 and V_1 are the internal energy, entropy and volume of phase (1) and U_2, S_2 and V_2 the same quantities for phase (2) then we can carry out an integration in which the system starts completely in phase (1) and ends up at the same temperature and pressure completely in phase (2). We then get:

$$U_2 - U_1 = T(S_2 - S_1) - p(V_2 - V_1)$$

or

$$U_2 - TS_2 + pV_2 = U_1 - TS_1 + pV_1 \tag{43}$$

The quantity $U - TS + pV$ is, as we have seen, the Gibbs free energy of the system. So we can deduce that if two phases can transform into each other at constant temperature and pressure, the Gibbs free energy of a given mass of one phase must equal the Gibbs free energy of the same mass of the other phase.

52

We have assumed that all this took place at temperature T and pressure p. There are, however, a succession of temperatures and pressures at which the two phases are in equilibrium. A diagram showing such values of p and T is called a phase diagram and a typical diagram is shown in Figure 5. We now consider a neighbouring temperature and pressure $T + dT$, $p + dp$ at which the two phases are still in equilibrium, i.e. at an adjacent point on the melting curve or on the vapour pressure curve. We must still have equality between the Gibbs free energies. At the original pressure and temperature p, T: $G_1 = G_2$; at the new pressure and temperature $G_1 + dG_1 = G_2 + dG_2$, i.e. $dG_1 = dG_2$ as we move along the equilibrium curve. Thus since

$$dG_1 = -S_1 \, dT + V_1 \, dp$$

and

$$dG_2 = -S_2 \, dT + V_2 \, dp \qquad \text{(see Table 3)}$$

$$-S_1 \, dT + V_1 \, dp = -S_2 \, dT + V_2 \, dp$$

or

$$\frac{dp}{dT} = \frac{S_2 - S_1}{V_2 - V_1} = \frac{\Delta S}{\Delta V} \qquad (44)$$

which is the Clausius–Clapeyron equation. It shows that the slope of the equilibrium curve is determined by the entropy change and the volume change which occur during the transition. Most solids absorb heat on melting and turn into liquids of larger volume so that both ΔS and ΔV are positive. This means, according to equation (44), that dp/dT is usually positive, i.e. that at higher pressures the melting temperature goes up. Water is unusual in this respect because when you melt ice (at least at fairly low pressures) the water formed has a *smaller* volume than the ice, i.e. ΔV is negative in this case. This means that the melting point of ice *decreases* when the pressure on it increases. This means that in Figure 5 the line separating the solid and liquid regions would have a negative slope.

The Joule–Thomson effect[16]

Suppose that a gas is forced at constant pressure p_1 through a porous plug from which it emerges at pressure p_2. Suppose that no heat enters or leaves the system. Now imagine that the volume of one gram-molecule (one mole for short) of the gas going in is V_1 and that of the gas coming out is V_2. Let the corresponding internal energies be U_1 and U_2. Now we will apply the first law of thermodynamics to the process in which one mole of gas passes through the plug.

$$\Delta U = U_2 - U_1 = W + Q$$

$Q = 0$ since the process is adiabatic; the work done in pushing the gas into the system is $p_1 \Delta V = p_1 V_1$, whereas the work done by the emerging gas is $p_2 V_2$. Thus $W = p_1 V_1 - p_2 V_2$ and so we have

$$U_2 - U_1 = p_1 V_1 - p_2 V_2$$

or

$$U_2 + p_2 V_2 = U_1 + p_1 V_1 \tag{45}$$

Since the enthalpy $H = U + pV$ we can write this result as:

$$H_2 = H_1 \tag{46}$$

i.e. the enthalpy of the gas is unchanged by the process. Incidentally, the same result would be true of any flow process under adiabatic conditions (for example, the flow of gas along an insulated pipe) in which p_1, V_1 and p_2, V_2 characterise the input and output conditions of the gas.

In general, the temperature of the gas will change in passing through the plug; this is the Joule–Thomson effect. If for a small pressure change δp, the temperature change is δT, the coefficient $(\delta T/\delta p)_H$ is called the Joule–Thomson coefficient of the gas under these conditions.

Let us now make use of some thermodynamic transformations to express this coefficient in terms of more readily measurable quantities.

We saw that in the Joule–Thomson process the quantity $U + pV$ was unchanged, i.e. $\delta(U + pV) = \delta U + p\,\delta V + V\,\delta p = 0$.

But from our equation (32)

$$\delta U + p\,\delta V = T\,\delta S$$

Therefore in the Joule–Thomson process

$$T\,\delta S + V\,\delta p = 0 \qquad \text{(cf. equation (37) above)}$$

This change in entropy δS can be re-expressed as due to the changes in T and p as follows:

$$\delta S = \left(\frac{\partial S}{\partial T}\right)_p \delta T + \left(\frac{\partial S}{\partial p}\right)_T \delta p$$

On substitution we get:

$$V\,\delta p = -T\left(\frac{\partial S}{\partial T}\right)_p \delta T - T\left(\frac{\partial S}{\partial p}\right)_T \delta p$$

By the Maxwell transformation:

$$\left(\frac{\partial S}{\partial p}\right)_T = -\left(\frac{\partial V}{\partial T}\right)_p$$

we have finally

$$\left(\frac{\delta T}{\delta p}\right)_H = \frac{T\left(\frac{\partial V}{\partial T}\right)_p - V}{C_p} \tag{47}$$

in which we have written $T(\partial S/\partial T)_p = C_p$.

Thus we have expressed the Joule–Thomson coefficient in terms of the volume of the gas, its heat capacity and its expansion coefficient, all of which are fairly easy to measure. This sort of manipulation in which we re-express a certain quantity in terms of others which may be more convenient is quite typical of one aspect of thermodynamics.

The Joule–Thomson process that we have just been considering is in fact an irreversible one; before concluding our discussion of the thermodynamic aspects of entropy we will look at some further examples of irreversible changes and their influence on entropy.

Exercises

Q1　1 kg of water at 20°C is converted into ice at −10°C all at atmospheric pressure. The heat capacity of water at constant pressure is 4200 J K^{-1} kg^{-1} and that of ice 2100 J K^{-1} kg^{-1}; the heat of fusion of ice at 0°C is 3.36×10^5 J kg^{-1}. What is the total change in entropy of the water–ice system?
If the density of water at 0°C is taken as 10% greater than that of ice, what is the slope of the melting curve of ice at this temperature? Give both sign and size.

Q2　A liquid obeys Trouton's rule, according to which $\lambda/RT = 10$, where λ is the latent heat of vaporisation at absolute temperature T and R is the gas constant per mole. By treating the vapour as a perfect gas and neglecting the volume of the liquid in comparison with that of the vapour, derive a relationship between the vapour pressure and the temperature of the liquid.

Q3　10^{-3} m^3 of lead is compressed reversibly and isothermally at room temperature from 1 to 1000 atmospheres pressure. Use one of Maxwell's thermodynamic relations to find (a) the change in entropy; (b) the heat given out; and (c) the change in internal energy of the lead. (Cf. also problems Q3 in Chapters 2 and 3.)
The isothermal compressibility of lead, $-V^{-1}(\partial V/\partial p)_T$, is 2.2×10^{-6} atm^{-1} and its volume coefficient of thermal expansion $V^{-1}(\partial V/\partial T)_p$ is 8×10^{-5} K^{-1}. Both can be assumed independent of pressure. Take 1 atm = 10^5 Pa.

Q4　Calculate the change in temperature of the lead in the adiabatic compression of Q3 Chapter 3.
(Hint: write down the partial differential coefficient of temperature with respect to pressure in an adiabatic, reversible process and convert this by a Maxwell transformation. Then re-express the result in terms of measurable quantities. Some of the data needed are given in the preceding problem Q3.)

In addition the value of C_p for lead at room temperature can be taken as $25 \, \text{J K}^{-1} \text{mole}^{-1}$ and the molar volume as $18.3 \times 10^{-6} \text{m}^3$.)

Q5 1 mole of a perfect gas for which C_V is $3R/2$, independent of temperature, is taken from a temperature of 100 K and pressure of 10^5 Pa to 400 K and 8×10^5 Pa by two different paths. (1) It goes at constant volume from 100 K to 400 K and then isothermally to the final pressure. (2) It goes at constant pressure from 100 K to 400 K and then isothermally to the final volume.

Calculate the heat absorbed or given out in each step and the algebraic sum for each path. Compare these with the corresponding entropy changes and show that the total entropy change is the same for each path ($\ln 2 = 0.693$).

6

Entropy in irreversible changes

Entropy is a thermodynamic variable which depends only on the state of the system and, because in any arbitrary process the entropy change depends *only* on the initial and final states, it now no longer matters whether or not the process is *reversible*. At first this seems paradoxical, but we must bear in mind that in *measuring* ΔS we are still effectively confined to reversible changes under well-defined conditions. Only *afterwards* when we can imagine that all states of the system have been mapped by means of these reversible changes, so that the value of S is known for each state, can we consider irreversible changes. Our situation is indeed quite similar to that encountered in the measurement of the internal energy function, U. To measure changes in U we were restricted to adiabatic paths along which $\Delta U = W$, the work done on the system. However, once the value of U in all states of the system is known, we can then make use of this knowledge in discussing non-adiabatic processes; in this way Q was defined. In a very similar manner we shall make use of S, which is measured by means of *reversible* processes, to discuss irreversible processes.

Heat flow

Whenever heat flows from a body at a high temperature to another at a lower temperature the process is irreversible. By the second law of thermodynamics the heat that has flowed into the body at the lower temperature cannot be recovered and transferred to the body at the higher temperature without the expenditure of work. It is therefore impossible to restore completely the situation that existed before the flow of heat. Associated with this process is an increase in entropy as we can see by the following argument.

57

Imagine that an amount of heat, Q, flows from a body at temperature T_1 to one at T_2. We can imagine that the bodies have sufficiently large heat capacities that their temperatures are effectively unchanged by the heat flow. Consequently the body at temperature T_1 loses an amount of entropy equal to Q/T_1 while the body at the lower temperature gains an amount of entropy of Q/T_2. If we can neglect any changes in the medium through which the heat flowed (this could be a small conducting wire), the net increase in entropy is therefore $Q/T_2 - Q/T_1$. Since heat always flows from a higher temperature to a lower one, T_1 is greater than T_2, so this quantity is always positive, i.e. the entropy must always increase.

If we want to make a transfer of heat which is very nearly reversible we must do so by means of a very small temperature difference. If in the example we have just discussed the difference between T_1 and T_2 is made quite small compared with either T_1 or T_2, we can write the entropy increase as

$$\delta S = Q\,\delta\left(\frac{1}{T}\right) = -Q\frac{\delta T}{T^2} \tag{48}$$

(Since the final temperature must be lower than the initial temperature, δT must be negative so that this quantity is always positive.) The amount of entropy transferred from one body to the other is equal to Q/T. If, therefore, we keep this quantity constant and wish to make the irreversible entropy increase as small as possible we must make $\delta T/T$ correspondingly small. For any given arrangement this means that the more nearly reversible the process is, the slower it has to be since the rate of heat flow depends upon δT. In principle, however, the process can be made as nearly reversible as we choose.

A numerical example will perhaps make clearer how the irreversible increase in entropy depends on the temperatures between which a heat transfer takes place. Suppose that we wish to cool a block of lead whose heat capacity is $1000\,\mathrm{J\,K^{-1}}$ from 200 K to 100 K. We shall assume that its heat capacity does not depend on the temperature. Let us first cool the lead by simply plunging it into a liquid bath at 100 K. What is the entropy change of the lead? What is the entropy change of the bath?

The entropy change of the metal is just

$$\int_{200}^{100} \frac{C}{T}\,\mathrm{d}T$$

where C is its heat capacity (cf. equation (24)). Thus

$$\Delta S_M = C\ln\frac{T_2}{T_1} = -1000\ln 2\ \mathrm{J\,K^{-1}} = -693\ \text{e.u.}$$

(for brevity let us replace $\mathrm{J\,K^{-1}}$ by entropy units, e.u. for short). The change in entropy of the bath is Q/T_B where Q is the heat received from the metal and T_B is the bath temperature. So $\Delta S_B = 100\,000/100\ \text{e.u.} = 1000\ \text{e.u.}$ The entropy of the metal thus decreases by 693 e.u. and the entropy of the bath increases by 1000 e.u. The net increase in entropy is therefore 307 e.u. in this process.

Now suppose we cool the lead block by cooling it first with a temperature bath at 150 K and then with the bath at 100 K. What is the net increase in entropy? The total entropy change of the lead (which depends only on its initial and final states) is unchanged and is equal to −693 e.u. However, the entropy increase of the baths is now altered. The bath at 150 K receives 50 000 J of heat from the metal so that its entropy increases by 50 000/150 e.u. = 333 e.u. The other bath receives the same amount of heat and so its entropy increases by 50 000/100 e.u. = 500 e.u. Thus the total increase in entropy of the two baths is now 833 e.u. and the net increase in entropy of the metal and baths together is now 140 e.u.

Suppose we now go a step further and use four temperature baths at 175, 150, 125 and 100 K. By the same sort of calculation it is easily shown that the entropy increase in the baths is now only 760 e.u. so that the net increase in entropy of baths and metal is now reduced to 67 e.u.

Obviously if we go on increasing the number of temperature baths the irreversible increase in entropy goes down until with a continuous range of such baths the increase in the entropy of the baths just equals the decrease in the entropy of the metal. The whole process would then be reversible.

Adiabatic processes which are irreversible

We have just seen that whenever there is an irreversible heat transfer from one body to another there is a net increase in entropy. We will now show that in any adiabatic process which is irreversible there is also always an increase in entropy. First let us take a particular case.

Consider a thermally insulated container divided by a partition into two parts each of volume V. One of these contains one mole of a gas which we will assume can be treated as an ideal gas. The other part is empty. We now want to see what happens to the entropy of the gas when the partition is removed and it is allowed to occupy the full volume of the container.

If the temperature of the gas initially is T, the entropy of the gas when confined to the volume V can be written down at once. It is

$$S_i = R \ln V + C_V \ln T + \text{constant}$$

C_V is the molar heat capacity at constant volume which we take to be independent of the temperature.

After the removal of the partition the gas expands into the other half of the container and when all turbulence has ceased the gas comes once more into equilibrium. In this expansion the gas has done no work and no heat has entered or left it so that its internal energy is unchanged. Moreover, we know from Joule's law that if the gas is an ideal one its temperature is

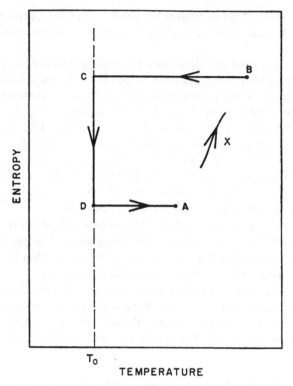

Figure 6 The initial change from state *A* to state *B* is by an irreversible process and therefore cannot be represented on the *S–T* diagram (which can show only quasi-static changes). It is indicated by the arrow *X*.

unchanged in this process. We can therefore write down the final entropy of the gas (which now occupies twice the original volume) as follows:

$$S_f = R \ln 2V + C_V \ln T + \text{constant}$$

Since the constant is the same in both expressions for the entropy, the entropy change is given by

$$\Delta S = S_f - S_i = R \ln 2$$

We therefore see that the entropy of the gas has *increased* even though no heat entered it or left it. At first sight this may seem puzzling because we define entropy in terms of the heat entering or leaving a system. The explanation is that the process we have just considered is *not a reversible one*. In irreversible processes the entropy *can* change even if the process is adiabatic, but in such changes the entropy *always increases*. We will now prove this in a more general way.

Consider a system whose initial state is represented by the point *A* on the entropy–temperature diagram (Figure 6) and suppose that it undergoes some irreversible adiabatic change and arrives at the state indicated by B in the diagram. (Since the states between A and B are not equilibrium states we cannot represent the path on this diagram.) We now want to show that the entropy of the state B must be higher than that of A. To do this we imagine that we have a heat reservoir at some temperature T_0. We now make our system, which is in the state B, undergo a reversible adiabatic change (i.e. at constant entropy) until it reaches the temperature T_0 (point C). At this temperature we make it undergo an isothermal change until its entropy has the value with which it started in state A. To do this, we must add heat to the system or take heat from it; let this quantity be Q. (Q is positive if heat enters the system and negative if it leaves it.) The system is now in a state represented by the point D. We now complete the cycle by making an adiabatic reversible change from the state D back to A.

We now have a complete cycle consisting of one isothermal process and three adiabatic processes of which one (between A and B) is irreversible. Since the system returns to its initial state its internal energy change ΔU is zero. If, therefore, the work done on the system during the cycle is W,

$$W + Q = 0$$

since the only heat entering or leaving the system is Q.

We therefore have a cycle in which our system either absorbs the heat, Q, and produces a corresponding amount of work, W; or it absorbs an amount of work, W, and gives out a corresponding amount of heat, Q. By the second law of thermodynamics only the second alternative is allowed. If the first were possible we should have a cyclic process which did nothing else but produce work from a single temperature source at T_0. This work could then obviously be used in some suitable irreversible process to deliver an equivalent amount of heat to another body at any temperature, even one above T_0. This, the second law asserts, is impossible. Our conclusion, therefore, is that only the second alternative is physically possible, namely that the work, W, is done on the system and the heat, Q, leaves the system. This in turn means that at the temperature T_0 the system *loses* the amount of heat, Q, and its entropy therefore diminishes by $\Delta S = Q/T_0$. Therefore the point D must be lower than the point C and therefore the point A must be lower than the point B. If the initial change were reversible A and B would have the same entropy. In all other cases the entropy of B must be greater than that of A. Therefore we have shown that whenever a system undergoes an irreversible adiabatic change between two equilibrium states its entropy must increase.

We thus see that the entropy of a system can change in two distinct ways: (i) there may be a heat flow (in or out) between the system and its surroundings; (ii) it may undergo an irreversible adiabatic change. Let the total change in entropy be dS and suppose that this is divided into two parts $(dS)_{ext}$ and $(dS)_{int}$. The first part $(dS)_{ext}$ arises from an exchange of heat, $đq$, with the surroundings and the second is due to any adiabatic irreversible change. We assume that the system has a uniform temperature T so that $(dS)_{ext} = đq/T$.

Thus

$$dS = \frac{đq}{T} + (dS)_{int}$$

However, as we have already seen, $(dS)_{int}$ which is due to irreversible adiabatic processes must always be positive. Therefore

$$dS \geq đq/T$$

The equality sign applies if there are no irreversible changes; otherwise the inequality holds. If the system is not at a uniform temperature, then we must divide it into portions which are and apply our result to each bit.

Now let us apply this result to a system which is thermally insulated so that $đq = 0$ and of fixed volume so that $đw = 0$. Then $dU = 0$ and $dV = 0$. Under these conditions

$$dS \geq 0$$

in any change. This means that *in any closed system whose volume and internal energy are fixed the entropy S tends to a maximum*, and when this is achieved all change ceases and the system is in equilibrium.

In concluding this section on entropy as a thermodynamic variable I would like to quote from Clausius' paper of 1865.[13] It was in this paper that he first used the name 'entropy' and first wrote down what I have called the central equation of thermodynamics (equation 32):

$$dU = T\,dS - p\,dV$$

After considering the entropy of a number of systems and what its physical significance is, Clausius ends the paper with the following paragraph:

> For the present I will limit myself to quoting the following result: if we imagine the same quantity, which in the case of a single body I have called its entropy, formed in a consistent manner for the whole universe (taking into account all the conditions), and if at the same time we use the other notion, energy, with its simpler meaning, we can formulate the fundamental laws of the universe corresponding to the two laws of the mechanical theory of heat in the following simple form:
>
> 1 The energy of the universe is constant.
> 2 The entropy of the universe tends to a maximum.

Exercises

Q1   A 100 ohm resistor is immersed in a constant temperature bath at 300 K. A current of 10 amps flows through the resistor and a steady state is achieved.

(a) Is there a flow of heat into the resistor? (b) Is there a flow of heat into the water? (c) Is work done on the resistor? (d) Apply the first law of thermodynamics to the resistor in its steady state. (e) Does the entropy of the resistor change? (f) Does the entropy of the environment change? (g) If so in either case, at what rate?

Q2   (a) 1 mole of oxygen at 300 K expands from a cylinder of volume 5 litres into an evacuated cylinder of the same volume (a so-called free expansion). When the temperature of the oxygen is again uniform at 300 K, what is the entropy change of the oxygen, which we assume can be treated as an ideal gas? What is the entropy change of the surroundings? (b) If the volume change from 5 to 10 litres took place reversibly and isothermally, what would be the entropy change of the oxygen? What would be the change in entropy of the heat bath used to keep the temperature constant?

PART TWO

The statistical interpretation of entropy

7

The statistical approach:
a specific example

We must now shift our point of view. So far we have been dealing with large-scale pieces of matter described in terms of large-scale measurements. Now we shall look at the behaviour of matter from a microscopic point of view, i.e. in terms of atoms and their behaviour. I have already pointed out that a complete mechanical description of the atoms that make up one gram of helium gas needs something like 10^{23} parameters whereas a thermodynamic description requires only 3. How then is it possible to handle such a huge number of variables and make sense out of them? A clue to the answer comes from the kinetic theory of gases. This theory allows us to calculate, for example, the pressure of a gas in a vessel from the average rate of change of momentum of the molecules striking unit area of the walls. Notice that we have to take an *average* and this is really the key to the problem. By suitably averaging certain properties of the individual atoms we are able to arrive at a *few* parameters which are enough to characterise the behaviour of the gas from the large-scale point of view. Here is another, perhaps more familiar, example. If you sit in the audience of a theatre when people are applauding, you can hear the individual handclaps of the people around you but if you listen to the applause from a long way off you hear only a rather uniform noise. This is really an average noise from the whole audience and the individual characteristics have been lost.

The kinetic theory of gases was already well developed by about the middle of the nineteenth century when the mechanical theory of heat became thoroughly established through the work of Joule. It was, perhaps, natural that people should try to link the two points of view, i.e. the thermodynamic and the mechanical. Boltzmann (1844–1906) was the first to succeed in interpreting entropy from a microscopic point of view. The kinetic theory of gases was well known to him; he developed it and produced from it a most powerful method of studying the thermodynamic properties of large numbers of similar molecules. He was in fact the founder of what we now call 'statistical mechanics'. The general ideas underlying the subject are implied by its name: that is to say, the individual atoms are assumed to obey the laws of *mechanics* but by adopting a suitable statistical approach we can deduce from their average behaviour properties corresponding to the *thermodynamic* properties of all the atoms together.[1]

A second important clue as to how to tackle the problem of statistical mechanics also comes from the kinetic theory of gases. This is the law of distribution of velocities among the molecules of a gas. This law was first formulated by Maxwell (1831–1879) in 1859 and formed the starting point of Boltzmann's researches into the statistical interpretation of entropy. The form of the distribution is illustrated in Figure 7 which shows how the number of molecules having velocities in a certain velocity range varies with velocity at a given temperature. At this point we shall not go into the derivation of this distribution law; we shall come to that later. At present all we need to notice is that such a distribution exists; more precisely, that when a gas is in equilibrium the numbers of molecules, having certain velocity ranges, do not change with time. Obviously the velocity of any particular molecule changes frequently – at each collision, in fact. However, these collisions, as Boltzmann showed for a number of different models, are such as to maintain the equilibrium distribution of velocities in the gas as a whole.

We can generalise this result. In any large-scale collection of molecules which have achieved thermodynamic equilibrium there is in addition to the velocity distribution a certain distribution of *energy* among the molecules which does not change with time (i.e. we can in principle specify a function analogous to that in Figure 7 but now giving the energy instead of the velocity of the molecules). This result cannot be so widely applied as that for the velocities for the following reason: we can only specify the total energy of a molecule if we can assign private, individual energies to each molecule. If the molecules are strongly interacting with each other there will be a mutual potential energy shared among several or many molecules and the individual molecule cannot be assigned a precise total energy. However, as long as we are dealing with *weakly interacting* molecules, we can expect to find a characteristic energy distribution among them when equilibrium is achieved in a large-scale collection of them. (Such a collection will be referred to as a collection of 'independent molecules'.)

We see therefore that in addition to the truly large-scale parameters such as pressure, volume and temperature which, at equilibrium, stay constant in time, there are these velocity and energy distributions which are also time-independent and are also characteristic parameters of the system (although giving more detail than p, V, T). As we shall see, the energy distribution at equilibrium will play an important part in our development of statistical mechanics and our first aim will be to find out exactly what this distribution is.

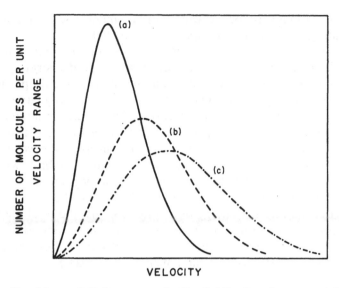

Figure 7 The Maxwell–Boltzmann velocity distribution in a gas at three different temperatures. The temperatures increase in the order (a) to (b) to (c).

Because the energy of atoms and molecules has such a prominent role in our discussions, I will mention here some of the forms of energy that atoms and molecules commonly possess.

In a gas the molecules obviously have the kinetic energy associated with their movement to and fro inside their container. This is often called their 'translational' motion and the energy that goes with it 'translational' kinetic energy. This energy will exist in all gases no matter what kind of atoms or molecules they contain.

In solids the mean position of the molecules is fixed (apart from diffusion) and their vibration about this mean position is an important kind of thermal motion; their energy then consists partly of kinetic and partly of potential energy. This kind of vibration occurs in all solids. In liquids the molecules have translational energy (since their mean positions are not fixed) but they may also have vibrational energy rather like the molecules in a solid.

In addition to these kinds of energy which are universal, molecules can have other kinds which depend on the nature of the specific molecule. For instance, a molecule of two atoms will be able to rotate and have rotational energy. The two atoms in the molecule will also be able to vibrate along the line joining them so that the molecule has internal vibrational energy. More complex molecules will have more complex modes of vibration. Another example: the electrons in an atom can have different energy states (e.g. the Bohr energy states in a hydrogen atom), so that atoms or molecules can have energy associated with excited electronic states. Here is a final example. If a molecule has a magnetic or electric dipole moment it will have magnetic or electrostatic energy in an appropriate applied field.

These are just some of the more important kinds of energy that molecules can have. We shall now see how some of these forms of energy, possessed by individual atoms or molecules, affect the thermodynamic properties (e.g. heat capacity or entropy) of systems formed from enormously large numbers of such molecules. (In referring to collections of molecules in large numbers I shall usually use the word 'assembly' to emphasise that it is made up of large numbers of individual molecules. This is in contrast to the word 'system' used in discussing thermodynamic properties where the emphasis was on the singleness of the entity.) We shall assume from the beginning – and this is a very important but very reasonable assumption – that the thermodynamic quantity U, i.e. the internal energy of a system, is just the total mechanical energy, potential and kinetic, of all the atoms and molecules composing it. (In the statistical treatment which follows I shall denote this energy by E simply to conform to the usage common in other books.)

A specific example

As our first introduction to the methods of statistical mechanics let us consider a specific and simple example. (This follows the lines of one given by Boltzmann in a paper of 1877 in which he explains very clearly his statistical interpretation of entropy.)[2]

Consider an assembly of seven identical atoms and suppose that each atom can take on only certain definite energies: 0, ε, 2ε, 3ε, 4ε etc. This limitation on the possible values of the energy of atoms or molecules is characteristic of quantum mechanics; in classical mechanics, of course, a particle would have a continuous range of energies.

It is, perhaps, of interest that Boltzmann, who was working long before the invention of quantum mechanics, used this discrete set of energies. At that time there was no justification for this and Boltzmann introduced it simply as a helpful device with the intention finally of allowing the value of ε to become vanishingly small, thus making the set of discrete levels go over to the continuum. Planck, too, used a similar approach in tackling the problem of radiation which led ultimately to his famous theory of quanta.

Returning to our assembly of seven atoms, we now fix the total energy of the assembly as a whole. Let this energy be 7ε. We also imagine that our atoms form a solid so that they cannot wander about. (Boltzmann discussed a gas, but a solid is in some ways rather simpler.) If you wish to have a definite model in mind, you can think of a row of seven atoms each of which can vibrate like a simple harmonic oscillator about its mean position. If the atoms can vibrate only in one dimension their energy levels will in fact form a simple arithmetical progression of the kind we have assumed. Alternatively, you can think of the atoms not as vibrating but as being dipoles and having equally spaced magnetic energy levels in an external field.

The total energy available to the assembly is 7ε and we imagine that this energy can be shared in all different possible ways among the atoms. In a gas this exchange of energy could be brought about by collisions; in our more artificial model we shall simply imagine that it is possible without saying precisely how. Note, however, that in this example and all the subsequent treatment we shall confine ourselves to weakly interacting atoms (or molecules) whose energy of interaction does not contribute appreciably to the total energy of the assembly. Since the maximum energy available to the assembly is 7ε, obviously no single atom can have more than this. For the present purposes, therefore, each atom can be in one of eight different energy levels:

$$0, \ \varepsilon, \ 2\varepsilon, \ 3\varepsilon, \ 4\varepsilon, \ 5\varepsilon, \ 6\varepsilon \text{ and } 7\varepsilon$$

As I have already mentioned, we wish to find out how the energy available to the assembly is distributed among the atoms at equilibrium. To do this we will first find out what possible distributions of energy are consistent with the given total energy. Take a typical distribution and assume that

n_0 atoms have energy 0
n_1 atoms have energy ε
n_2 atoms have energy 2ε
$\ldots$
n_7 atoms have energy 7ε (49)

We require that

$$n_0 + n_1 + n_2 + \ldots + n_7 = 7 \tag{50}$$

and

$$\varepsilon n_1 + 2\varepsilon n_2 + 3\varepsilon n_3 + \ldots + 7\varepsilon n_7 = 7\varepsilon \tag{51}$$

in order that the total number of atoms be 7 and their total energy 7ε. If these conditions are satisfied the n's define a *possible* distribution of energy in our assembly.

We wish now to do two things:

(i) To write down all the possible distribution numbers of the kind indicated in (49) above which are consistent with the restrictions implied by equations (50) and (51).

(ii) To evaluate how many different atomic arrangements correspond to each of these distributions, i.e. to work out in how many distinguishably different ways the energy can be given to the atoms for each set of n's. (The examples below will make this clear.)

Notice carefully the difference between (i) and (ii); this is very important.

In (i) we want to know about the distribution numbers and we do not care which particular atoms have a particular energy, only *how many*. In (ii), the identity of the atoms having a certain energy *is* important. We imagine the atoms to be labelled (1 to 7, say). Then, if in one atomic arrangement, the atom numbered 5 has all the energy (7ε) and the rest none, this is to be counted as distinguishable from the arrangement in which the atom numbered 3 has all the energy and the others none. An arrangement in which we specify exactly which atoms have which particular energy we shall call a microscopic state of the assembly (microstate, for short). In (ii) therefore we wish to count up how many distinguishable microstates correspond to each set of distribution numbers, n_0, n_1, n_2 etc. We shall then use this information to decide which energy distribution (i.e. which set of distribution numbers) is the most probable and identify this with the energy distribution at equilibrium.

Task (i): Possible distribution numbers

In our present problem there are eight distribution numbers, n_0, n_1, n_2 etc., corresponding to the eight possible energy levels (listed in (49)). Take first a very simple distribution of energy: in this, one atom (it doesn't matter which) has all the energy (7ε) and the other six have none. Thus $n_7 = 1$ and $n_0 = 6$. All the other numbers $n_1 = n_2 = n_3 = n_4 = n_5 = n_6$ are zero. We could write this distribution symbolically as 60000001 corresponding to the sequence $n_0 n_1 n_2 n_3 n_4 n_5 n_6 n_7$.

A second possible distribution of energy is one in which one atom has energy 6ε and one atom has ε. This means that $n_1 = n_6 = 1$, $n_0 = 5$ and the other ns are zero; i.e. this distribution can be written as 51000010.

In another possible distribution one atom has energy 5ε and two atoms each have ε, the rest having zero; i.e. 42000100.

Note that in all these distributions $n_0 + n_1 + n_2 + n_3 + n_4 + n_5 + n_6 + n_7 = 7$, and $n_1\varepsilon + 2n_2\varepsilon + \ldots + 7n_7\varepsilon = 7\varepsilon$ as the conditions of the problem demand. These and the other possible distributions (making a total of fifteen possible ones in all) are listed in Table 4. You should check one or two of them to make sure that they correspond to the correct total energy of the whole system.

This completes task (i).

Task (ii): Counting the microstates

Consider first the energy distribution which is numbered 1 in Table 4 (these distributions are numbered in the left-hand column). In this distribution one atom has all the energy and the other six have none. In specifying the microstates corresponding to this distribution we can say that each atom in turn may have all the energy so that there are seven distinguishable

Table 4 Possible distribution numbers

	n_0	n_1	n_2	n_3	n_4	n_5	n_6	n_7	Number of microstates P
1	6							1	7
2	5	1					1		42
3	5		1			1			42
4	5			1	1				42
5	4	2				1			105
6	4	1	1		1				210
7	4	1		2					105
8	4		2	1					105
9	3	3			1				140
10	3	2	1	1					420
11	3	1	3						140
12	2	4		1					105
13	2	3	2						210
14	1	5	1						42
15		7							1

Table 5 Microscopic states corresponding to distribution number 1

Atom number	1	2	3	4	5	6	7
Microscopic state							
1	7ε	0	0	0	0	0	0
2	0	7ε	0	0	0	0	0
3	0	0	7ε	0	0	0	0
4	0	0	0	7ε	0	0	0
5	0	0	0	0	7ε	0	0
6	0	0	0	0	0	7ε	0
7	0	0	0	0	0	0	7ε

microscopic states of the system corresponding to distribution number 1. These seven are shown in Table 5.

In this table the atoms have been numbered from 1 to 7. Underneath, in each horizontal row, are shown the energies that each atom has in each of the seven different microscopic states. Now the identity of the atom which has the energy is important.

Distribution number 2

When we come to the second set of distribution numbers (number 2 of Table 4) the number of distinguishable microstates is already so large that they

Table 6 Some possible microstates of distribution number 2

Atom number	1	2	3	4	5	6	7
Microstate							
1	6ε	ε	0	0	0	0	0
2	6ε	0	ε	0	0	0	0
3	6ε	0	0	ε	0	0	0
....							
7	ε	6ε	0	0	0	0	0
8	0	6ε	ε	0	0	0	0
9	0	6ε	0	ε	0	0	0
....							
13	ε	0	6ε	0	0	0	0
....							

cannot all be tabulated separately as in Table 5. A few examples of the microstates will suffice to show how they are made up. Now, five atoms have zero energy, one has ε, and one has 6ε. Some possible microstates are shown in Table 6. In distributing the energy to the atoms so that one has 6ε, one has ε and the rest zero, first give 6ε to number 1 and ε to number 2; the rest get zero (microstate 1 of Table 6). Then still giving 6ε to number 1, now give ε to number 3, the rest having zero (microstate 2 of Table 6). Then give ε to numbers 4, 5, 6 and 7 in turn, thereby building up microstates 3, 4, 5 and 6 (only the first three are shown explicitly in the table to save space).

These six microstates are all that can be made with atom 1 having 6ε. Now give the 6ε to atom 2 and begin distributing the energy ε to atoms 1, 3, 4, 5, 6 and 7 in turn to create six new microstates. Only the first three of these microstates are shown in the table. Then give the 6ε to atom 3 and so on.

Thus for each atom in turn which has energy 6ε, there are six different ways of distributing the remaining energy ε to the other atoms. The atom which is to receive 6ε can be chosen in seven ways (i.e. each of the seven atoms in turn) so that the *total* number of microstates corresponding to this distribution of energy is $7 \times 6 = 42$.

Distribution numbers 3 and 4

These are essentially the same as the distribution number 2 which we have just discussed. Each has 42 microstates.

Distribution number 5

The counting of the microstates needs a little more care in this case. Some examples are shown in Table 7. In this distribution, one atom has energy

Table 7 Some possible microstates of distribution number 5

Atom number	1	2	3	4	5	6	7
Group 1	5ε	ε	ε				
	5ε	ε		ε			
	5ε	ε			ε		
Group 2	5ε		ε	ε			
	5ε		ε		ε		
	5ε		ε			ε	
Group 3	5ε			ε	ε		
	5ε			ε		ε	
	5ε			ε			ε
Group 4	5ε				ε	ε	
	5ε				ε		ε
Group 5	5ε					ε	ε

5ε, two atoms have ε and four have zero. It is quite easy to write down systematically all possible microstates and the only point of difference from the previous distributions is that now *two* atoms have the same energy ε. We have to be careful not to count the same microstate twice in different contexts.

For example, in the group 1 microstates of Table 7, atom 1 always has energy 5ε and atom 2 always has energy ε; then the remaining energy ε is given to the other atoms 3, 4, 5, 6 and 7 in turn.

In the group 2 microstates atom 1 always has energy 5ε and atom 3 always has energy ε; the remaining energy ε is then given to atoms 4, 5, 6 and 7 in turn. However, notice, *not* to atom 2 as this microstate has already been counted in group 1. Bearing this point in mind, we can then easily show that the total number of microstates corresponding to this distribution is 105.

Now we have seen how to write down the microstates systematically for any given set of distribution numbers n_0, n_1, n_2 etc. However, we do not want to know what the individual microstates are but only how many there are for each set of n's. What we want next, therefore, is a quick way of counting them.

The number of microstates corresponding to a given set of distribution numbers

In the present example, our problem is: in how many distinguishable ways can we distribute seven atoms so that n_0 go in the energy state zero, n_1 in

ε, n_2 in 2ε and so on? Obviously the actual values of the energy do not matter at this stage of the problem. We could just as well ask: in how many distinguishable ways can we put seven labelled objects in eight different boxes so that n_0 go in the first box, n_1 in the second box, n_2 in the third box and so on? Notice, however, that when we rephrase the question in this way, the *order* in which the atoms go in the box does not matter. When we say that two atoms are in box 3 we mean only that two atoms have the same energy (appropriate to that box) – there is no question of order or sequence here.

So we can rephrase the question yet again (to reduce it to a standard problem in permutations): in how many ways can we divide N labelled objects into heaps of n_0, n_1, n_2, ..., n_k, where $n_0 + n_1 + n_2 + \ldots + n_k = N$, with no particular order in each heap? (We have made the problem general by taking N objects instead of 8 because it is just as easy and we want the more general result.)

The number of such ways, which we will call P, is given by[3]

$$P = \frac{N!}{n_0! n_1! \ldots n_k!} \tag{52}$$

(Incidentally, 0! is counted as 1.)

By means of this formula we can now easily calculate the number of microstates corresponding to the remaining distribution numbers of Table 4. Moreover, we can also verify that the numbers we got by directly counting the states are correct. The whole sequence of numbers is listed in the last column of Table 4.

If you look at Table 4 you will see that to some sets of distribution numbers there correspond very many more microstates than to others. To take the extreme case, distribution number 15 of the table can be achieved in only one way, namely with each atom having energy ε. On the other hand, there are 420 microstates corresponding to the distribution number 10. What is the significance of this?

Boltzmann suggested that if you could observe such an assembly over a long period of time each microstate would occur with equal probability and you would find that any particular set of distribution numbers would occur in proportion to the number of microstates corresponding to that set. In terms of probability, we could say that the probability of the distribution number 6 (of Table 4) compared to that of number 14 would be in the ratio of 210 to 42 and so on from the others. The total number of microstates for all possible sets of distribution numbers is (if you add up the last column of the table) 1716. Thus the probability of finding the different sets of distribution numbers in turn would be (starting with no. 1) $\frac{7}{1716}$, $\frac{42}{1716}$, $\frac{42}{1716}$ . . ..

In the present example the assembly would be found most often in the state corresponding to the distribution number 10 since this has the biggest number of microstates (420) corresponding to it. If we take the same set of energy levels, but now with a very much larger number of molecules, N, and a correspondingly increased total energy, E, we would find that now the most probable distribution (i.e. that with the largest value of P) and those close to it have a much greater probability relative to the other distributions; if the number N approaches that of a large-scale physical system ($N \sim 10^{23}$) this probability becomes overwhelming.

This is the basic idea of statistical mechanics. First we specify the large-scale limitations of the assembly: i.e. we fix N, the number of molecules; we fix E, the total energy; and we fix the volume V available to the assembly. In principle we then know the energy levels available to the particles in the system from the laws of mechanics. We next consider all possible sets of distribution numbers and calculate the number of microstates, P, corresponding to each. We then pick out that set of distribution numbers for which P is a maximum. We may think of this as representing the state close to which the system will almost always be found if left to itself. That is, this corresponds to the state of thermodynamic equilibrium.

Already this idea of what thermodynamic equilibrium means in terms of our mechanical model gives us a powerful insight into the meaning of the second law of thermodynamics. If we are right in thinking of a thermodynamic system as fundamentally a mechanical one comprising an unimaginably large number of atoms or molecules, and if further such a mechanical system samples an impartial selection of all the microstates available to it, we can at once understand the irreversible nature of spontaneous changes in thermodynamics. Thermodynamic equilibrium, we say, corresponds to those states of the system for which P is a maximum or close to it. Moreover, in large-scale systems the number of such microstates is overwhelmingly large compared to that for all non-equilibrium states. If therefore we start with an isolated system in a state well away from equilibrium it must then be one to which comparatively few microstates correspond. When left to itself therefore the system will (with overwhelming probability) move towards equilibrium just because all microstates have equal probability and because an overwhelming number of them correspond to the equilibrium state.

Once in equilibrium, the system will stay there. In principle it could spontaneously move away from equilibrium, but in a large-scale system the odds against this are so enormous that the possibility can be ignored (see Chapter 10).

The model suggests a still further step. We see that at equilibrium P has its maximum value or a value close to it. We also know from thermodynamics that, in a closed and isolated system at equilibrium, the entropy function S likewise has a maximum. This suggests that S and P may be closely linked. As we shall see this is indeed true. This relationship between the entropy of a thermodynamic system at equilibrium and the number of microstates to which the system has access was Boltzmann's fundamental discovery; it lies at the heart of statistical mechanics.

In the next chapter we shall develop these ideas, first by working out the equilibrium distribution numbers in a more general way and then by using these results to show how the maximum value of P is related to the entropy.

Exercises

Q1 Construct a table analogous to Table 4 (p. 73) for an assembly of three atoms fixed in position with energy levels $0, \varepsilon, 2\varepsilon, 3\varepsilon$ and a total energy of the assembly of 3ε. What is the least and what is the most probable distribution of energy? What is the average distribution of energy?

Q2 In anticipation of the next chapter, let us look at Lagrange's method of undetermined multipliers and use it to solve the following two simultaneous equations:

$$x + y = 3 \quad \text{and} \quad x - y = 1.$$

These can be rewritten and labelled:

$$x + y - 3 = 0 \ \text{(a)} \quad \text{and} \quad x - y - 1 = 0 \ \text{(b)}$$

Now multiply (b) by an undetermined multiplier λ and add it to (a) to give:

$$x + y - 3 + \lambda(x - y - 1) = 0 \ \text{(c)}$$

If now (a) and (b) are true, (c) must also be true for any value of λ. Therefore separate out in (c) the terms in x, the terms in y and those independent of both. Now we can give λ any value we wish, so choose the values of λ first to make the coefficient of x vanish and then the coefficient of y vanish. This solves the equation, as you may check.

8

General ideas and development: the definition of a microstate

So far we have seen in outline the way in which we hope to relate the *mechanical* behaviour of individual molecules with their *thermodynamic* behaviour in large numbers. However, there are some important questions to answer before we work out the details of the programme. Why, you may ask, do we give equal weight to each distinguishable microscopic state of the assembly? We have counted them up and treated each distinguishable microstate impartially. Is this justified? The definition of a microstate of the assembly depends (a) on how we count the allowable states of the individual atom (e.g. are all the energy levels to be given equal weight in the first place?), and (b) on what we mean by the word 'distinguishable'.

Consider first (a). We have a choice here; we could have chosen our original states not in terms of energy at all but in terms of, for example, momentum. We define our states in terms of energy because we can then easily impose our restriction on the total energy of the assembly; nevertheless we may well have to assign a different importance to different energy levels.

As regards (b) we have taken the atoms as labelled and thus as distinguishable from each other. In terms of classical mechanics we might regard this as reasonable because, in principle, we can follow the path of an atom continuously and so give it a continuous identity. In quantum mechanics we cannot do this even in principle.

We see therefore that the methods of defining and counting the microstates of our assembly are *not* unambiguous and we are obliged to make a choice. Boltzmann, for example, found that he had to assign equal weights to equal ranges of a molecule's *momentum* (*not* its *energy*) to ensure that the average properties of his collection of molecules corresponded with the thermodynamic properties of an ideal gas.[4] When quantum mechanics superseded classical mechanics in the description of the mechanics of atoms, it became possible to define a microstate more simply and more convincingly in terms of the new mechanics. Nevertheless the definition of a microstate and the method of counting such states were chosen in the first place so that predictions based on them agreed with experiment. In short, we cannot prove, *a priori*, what the correct definition of a 'distinguishable microstate' is. Rather we make an assumption and test its consequences.

Right at the beginning therefore we make these two assumptions:

(i)   That the distinguishable microstates of an assembly of given E, N and V are the quantum mechanical stationary states corresponding to those values of E, N, V.

(ii)   That all such states are to be considered equally probable.

What do we mean by 'quantum mechanical stationary states'? The results which I am about to quote are taken over directly from quantum mechanical calculations. The important feature of them from the present point of view is that they show that an atom or electron or other microscopic entity can exist *only in certain definite and discrete energy states*. This result and the actual values of these energy states in certain simple cases are all that we need to know about quantum theory for our present purposes. On the other hand, this difference between quantum mechanics and classical mechanics (which allows a continuous range of energy values) is vital to an understanding of low temperature phenomena. Now let me give some simple examples of the stationary states of a *single* mechanical system, not an assembly of particles.

Consider a one-dimensional harmonic oscillator whose classical characteristic frequency is ν. According to quantum mechanics, the stationary states of such an oscillator have energies

$$\tfrac{1}{2}h\nu, \tfrac{3}{2}h\nu, \tfrac{5}{2}h\nu, \ldots, (n + \tfrac{1}{2})h\nu, \ldots$$

where n is any positive integer and h is Planck's constant. To each of these energies there belongs one and only one stationary state. (This result is essentially the same as that originally postulated by Planck, namely that the energy of a Hertzian oscillator of frequency ν could assume only the discrete set of values $nh\nu$ where n is an integer. The difference between these two results is that in the first case the lowest state has energy $\tfrac{1}{2}h\nu$ – the so-called 'zero-point' energy – while in the second this energy is zero.)

A rigid linear rotator, whose moment of inertia about a line perpendicular to its axis is I, has stationary states with energies given by

$$\varepsilon_j = \frac{h^2}{8\pi^2 I} j(j + 1)$$

where j is a positive integer. To each of these energies correspond $(2j + 1)$ stationary states. In this example, therefore, the number of stationary states varies with the energy of the state. In the language of quantum mechanics such energy levels are said to be 'degenerate'. In this case the levels are $(2j + 1)$-fold degenerate.

As a final example, consider the stationary states of a point particle of mass, m, confined to a cubical box of side L. This example will be very useful to us later when we come to consider the statistical mechanics of gases. The stationary states of such a particle have energies given by

$$\varepsilon_{p,q,r} = \frac{h^2}{2mL^2}(p^2 + q^2 + r^2)$$

where p, q and r are any positive or negative integers. These energy levels are non-degenerate.

Quantum mechanical calculations therefore give us the basic information we need about the stationary states that the individual atom or molecule can have. For example, a molecule of two atoms can vibrate and rotate and the whole molecule can move about; it will thus have, among others, stationary states of the kind just described. Once we know the stationary states that the individual molecules in our assembly can have, then the stationary states of the whole assembly are just suitable combinations of these individual stationary states. By 'suitable' here, I mean those whose total energy is equal to that prescribed for the assembly.

According to the point of view of quantum mechanics, molecules or atoms cannot even in principle be continuously observed so that they cannot maintain a continuous identity. Consequently molecules or atoms of the same kind in an assembly are to be thought of as fundamentally *indistinguishable*. In counting the distinguishable microstates of an assembly this must be borne in mind.

In what follows we shall deal with two kinds of assembly: one we shall use for discussing solids and the other for discussing gases. In a solid in which there is no diffusion each atom or molecule is on the average fixed in position, i.e. on its site in the lattice. In the solid-like assembly, therefore, the state of each atom or molecule is characterised not only by its quantum energy level (of which we have already given examples), but also by its position in the solid.

By contrast, the particles in a gas are to be thought of as completely *non-localised*. In a gas-like assembly each particle in a definite quantum state is to be considered as spread throughout the volume available to the gas. No particle therefore has a particular position and each particle of the gas is characterised *only* by its quantum energy level.

In brief, therefore, the particles in a solid-like assembly are characterised by both position and energy state: those in a gas-like assembly by their energy state only. For this reason the counting of the distinguishable microstates of these two kinds of assembly is different and we shall deal with them separately. Because the counting is somewhat easier, we shall first deal with the solid-like assembly; this is often referred to as an assembly of *localised* particles or elements.

More general treatment of localised assembly

For simplicity we shall assume that the energy levels available to each molecule are $\varepsilon_1, \varepsilon_2, \varepsilon_3, \ldots, \varepsilon_i, \ldots$ and that to each level there corresponds only one stationary state (i.e. each level is non-degenerate); it is not difficult to remove this limitation. Assume we have N molecules and that the total energy of the assembly is E. The volume of the assembly is fixed and equal to V. In general the energy levels $\varepsilon_1, \varepsilon_2$ etc. depend on the volume per molecule (V/N) and since both N and V are fixed so are the energy levels. Suppose that the energy is distributed so that

n_1 molecules have energy ε_1
n_2 molecules have energy ε_2
$\ldots$
n_i molecules have energy ε_i
$\ldots$

We now wish to compute the number of different stationary states of the assembly corresponding to this distribution of energy. Since the molecules can be thought of as effectively identified by the sites on which they are located (because they are on the average fixed in position) the number of different stationary states of the assembly is simply the number of distinguishable ways of arranging the molecules in the energy levels, i.e. it is the same problem as we discussed before. The answer is thus

$$P = \frac{N!}{n_1! n_2! \ldots n_i! \ldots} \tag{53}$$

According to Boltzmann's hypothesis, we must now look for those values of the distribution numbers (the n_i's) which make P a maximum. This and the distributions close to it will then, we believe, correspond to the state of thermodynamic equilibrium of our assembly. However, of course, we must also satisfy the restrictive conditions that

$$n_1 + n_2 + \ldots + n_i + \ldots = N: \quad \text{in brief} \quad \sum_i n_i = N,$$

and

$$n_1\varepsilon_1 + n_2\varepsilon_2 + \ldots + n_i\varepsilon_i + \ldots = E; \quad \text{in brief} \quad \sum_i n_i\varepsilon_i = E$$

To find the maximum value of P subject to these conditions requires some mathematical manipulation. Since P involves the products of $n_i!$, $n_2!$ etc., it is easier to work not with P but with $\ln P$ ($\ln P$ is short for $\log_e P$ – the natural logarithm of P). A maximum of $\ln P$ will also correspond to a maximum of P itself. Moreover, we want a simple approximation for the $n_i!$'s. For this we use Stirling's approximation. According to this

$$\ln n_i! \simeq n_i \ln n_i - n_i \tag{54}$$

and the approximation is a very good one if the n_i's are very large – as we assume they are in all the examples we deal with. Because the n_i's are large we can also treat them as being effectively continuous variables rather than integers.

First we write

$$\ln P = \ln N! - \ln n_1! - \ln n_2! - \ldots - \ln n_i! - \ldots$$

To find the maximum value of $\ln P$ we allow all the n_i's to change by small arbitrary amounts δn_i. The corresponding changes in $\ln P$, $\ln N!$, $\ln n_1!$ etc. we write as $\delta(\ln P)$, $\delta(\ln n_i!)$, $\delta(\ln N!)$ and so on. Obviously if N is fixed $\delta(\ln N!)$ is zero. So we have

$$\delta(\ln P) = -\delta(\ln n_1!) - \delta(\ln n_2!) - \ldots - \delta(\ln n_i!) - \ldots = -\sum_i \delta(\ln n_i!) \tag{55}$$

From the restrictive conditions we know that, in addition,

$$\delta n_1 + \delta n_2 + \ldots + \delta n_i + \ldots = \sum_i \delta n_i = 0 \tag{56}$$

$$\varepsilon_1 \delta n_1 + \varepsilon_2 \delta n_2 + \ldots + \varepsilon_i \delta n_i + \ldots = \sum_i \varepsilon_i \delta n_i = 0 \tag{57}$$

because E and N are fixed.

At a maximum (or a turning point) of $\ln P$, $\delta \ln P$ must vanish for any small changes in the n_i's so that we must have:

$$\sum_i \delta(\ln n_i!) = 0 \qquad\qquad \text{subject to (56) and (57)}$$

But

$$\delta \ln n_i! = \frac{\partial}{\partial n_i} (\ln n_i!) \, \delta n_i$$

$$= \frac{\partial}{\partial n_i} (n_i \ln n_i - n_i) \, \delta n_i \qquad\qquad \text{(from Stirling's formula)}$$

$$= \left(\ln n_i + \frac{n_i}{n_i} - 1 \right) \delta n_i$$

$$= \ln n_i \, \delta n_i$$

So that we have

$$\sum_i \ln n_i \, \delta n_i = 0 \tag{58}$$

subject to (56) and (57).

To take account of the restrictive conditions we could use them to eliminate two of the δn_i's (say δn_1 and δn_2) from equation (58). Instead we keep the symmetry of the expressions and use Lagrange's method of undetermined multipliers. Multiply equation (56) by the parameter α and (57) by the parameter β and add them to (58). We then get

$$(\alpha\, \delta n_1 + \beta\varepsilon_1\, \delta n_1 + \ln n_1\, \delta n_1) + (\alpha\, \delta n_2 + \beta\varepsilon_2\, \delta n_2 + \ln n_2\, \delta n_2)$$
$$+ \ldots + (\alpha\, \delta n_i + \beta\varepsilon_i\, \delta n_i + \ln n_i\, \delta n_i) + \ldots = 0$$

or

$$\sum_i (\alpha + \beta\varepsilon_i + \ln n_i)\, \delta n_i = 0 \qquad (59)$$

Instead of eliminating two of the δn_i's by means of the restrictive condition, we have introduced the parameters α and β which are now at our disposal. We may imagine that we now choose the values of α and β to make the first two brackets vanish. The remaining δn_i's are then independent so that to make the sum vanish for arbitrary values of these δn_i's, *all* the brackets must be zero. We therefore have quite generally that to make $\ln P$ a maximum

$$\alpha + \beta\varepsilon_i + \ln n_i = 0$$

or

$$n_i^* = e^{-\alpha - \beta\varepsilon_i} \qquad (60)$$

(It can be shown by a second variation that these values of the n_i's do indeed correspond to a maximum of P and not a minimum or point of inflexion.) This particular set of distribution numbers (denoted by n_i^*) thus correspond to the equilibrium distribution numbers of our assembly. This is then the characteristic energy distribution at equilibrium which we have been seeking.

The Boltzmann distribution

Now we would like to know something about the parameters α and β.
Since $\Sigma n_i^* = N$ we can readily eliminate α.

$$\sum_i n_i^* = \sum_i e^{-\alpha - \beta\varepsilon_i} = e^{-\alpha} \sum_i e^{-\beta\varepsilon_i} = N$$

Therefore

$$\frac{n_j^*}{N} = \frac{e^{-\beta\varepsilon_j}}{\displaystyle\sum_i e^{-\beta\varepsilon_i}} \qquad (61)$$

Notice that the denominator on the right-hand side does not depend on j since it is a sum over all the energy levels. The sum is a characteristic of the whole assembly not of any particular energy level.

Instead of attempting to eliminate β, we keep it in because it turns out to have a simple and important physical meaning. What do we know at this stage about β? To answer this let us first write down the restrictive condition on the energy of the assembly:

$$\sum n_i^* \varepsilon_i = E$$

Now substitute for the n_i^* from equation (61). Then:

$$\frac{\sum_i \varepsilon_i e^{-\beta\varepsilon_i}}{\sum_i e^{-\beta\varepsilon_i}} = \frac{E}{N}$$

Suppose now that N and V are fixed but that E is allowed to change. Because the volume V is fixed the ε_i's are all fixed so the only two variables in the equation are E and β. So if E changes, β must change and *vice versa*. Indeed, we can regard this equation as determining β in terms of E or, preferably, E in terms of β since it is in fact easier to see what happens to E when β changes than the other way round.

Consider what happens if β tends to infinity. Obviously all the terms in both sums become very small (because they are all of the form $A\,e^{-x}$ where x is now a very large number) and the biggest term in both the numerator and denominator will be the one with the smallest value of ε, i.e. the first term in each sum which thus involves ε_1, the energy of the lowest state. Therefore as $\beta \to \infty$

$$\frac{E}{N} \to \frac{\varepsilon_1 e^{-\beta\varepsilon_1}}{e^{-\beta\varepsilon_1}} = \varepsilon_1$$

So as $\beta \to \infty$, $E \to N\varepsilon_1$

This simply tells us that as $\beta \to \infty$, the total energy of the assembly takes on its lowest value, i.e. that corresponding to having all the molecules in their lowest energy states, ε_1.

Now suppose $\beta \to 0$. Then all the exponential terms tend towards unity and we have under these circumstances

$$\frac{E}{N} \to \frac{\sum_i \varepsilon_i}{\sum_i 1}$$

The numerator is thus just the sum of the energy values of all the occupied levels and the denominator is just the total number of occupied levels. If there is no limit to the values of the ε_i's, i.e. if there are an unlimited number of levels with ever-increasing energy, then the energy of the assembly will tend to infinity as $\beta \to 0$.

This then gives us some idea of what influence β has: if β is large the energy of the assembly must be small; if β is small the energy of the assembly must be large. In the limit as β tends to infinity the energy of the assembly takes on its minimum value; as β tends to zero, the energy takes on its maximum value. This suggests that β is inversely related to the absolute temperature of the assembly, T.

This, of course, is just a plausibility argument because so far we have not even shown that the assembly has any property analogous to temperature. However, in Chapter 9 we take up this question in detail and we shall then demonstrate that β has indeed the property of a temperature and moreover that $\beta = 1/kT$, where k is a constant. It will then also appear that k does not depend on the kind of assembly we are discussing and is in fact an important physical constant, now known as Boltzmann's constant. As we shall show later when we discuss the statistical mechanics of gases, $N_0 k = R$, where N_0 is Avogadro's number and R is the gas constant per mole. For the present we shall assume that we can write $\beta = 1/kT$ and shall make use of this result right away; it enables us to discuss the equilibrium distribution of energy among the molecules of our assembly in terms of *temperature* instead of in terms of E, the total *energy* of the assembly. Both physically and mathematically, this greatly clarifies and simplifies the discussion.

We can now rewrite equation (61):

$$\frac{n_j^*}{N} = \frac{e^{-\varepsilon_j/kT}}{\sum_i e^{-\varepsilon_i/kT}} \tag{62}$$

This expression then tells us how the number of molecules in each energy state varies with the energy of that state. To see more clearly what this means let us compare n_j^*, the number of molecules with energy ε_j, with n_l^*, the number with energy ε_l. Obviously:

$$\frac{n_l^*}{N} = \frac{e^{-\varepsilon_l/kT}}{\sum_i e^{-\varepsilon_i/kT}}$$

The ratio of the two numbers is then

$$\frac{n_j^*}{n_l^*} = \frac{e^{-\varepsilon_j/kT}}{e^{-\varepsilon_l/kT}} = e^{-(\varepsilon_j - \varepsilon_l)/kT} \tag{63}$$

Let us see what this implies. First, notice that only the energy *difference* $\Delta\varepsilon = \varepsilon_j - \varepsilon_l$ comes into the result so that provided that both energies are measured from the same origin it does not matter what that origin is. Second, and this is very important, what matters in determining the ratio of the occupation numbers is the ratio of the energy difference $\Delta\varepsilon$ to kT. Suppose that $\varepsilon_j = kT$ and $\varepsilon_l = 10kT$. Then $(\varepsilon_j - \varepsilon_l)/kT = -9$:

$$\frac{n_j^*}{n_l^*} = e^9 \simeq 8000$$

In this example, therefore, there are 8000 times more molecules in the lower energy state (ε_j) than in the upper. This brings out the important point that for a given temperature, many more molecules will be found in the lower energy states than in the higher; moreover, if this ratio is to be large, the difference in energy, $\Delta\varepsilon$, between the levels must be large compared to kT. Again and again we shall find the importance of kT as a standard of comparison for the differences in the energy levels of atoms or molecules.

The lowest energy level (call it ε_0) available to a molecule (or system) is often referred to as the 'ground state'. Suppose that in our present example we measure all energies relative to ε_0, and that n_0 is the number of molecules in this state. If n is the number of molecules with energy ε (above the ground state),

$$\frac{n}{n_0} = e^{-\varepsilon/kT} \tag{64}$$

for our assembly at temperature T. Such an energy distribution is frequently encountered in physical problems and, when proper account is taken of any degeneracy of the energy levels involved, is often referred to as a 'Boltzmann distribution'. The term $e^{-\varepsilon/kT}$ is also often called the 'Boltzmann factor'.

Let us now regard ε as a continuous variable and see how n/n_0 varies with ε at a given temperature; we will plot ε along the y-axis and n/n_0 along the x-axis. In fact, to make the curve valid for all temperatures, we shall plot not ε but ε/kT. This is shown in Figure 8. It shows how the occupation numbers fall off as the energy of the state increases; the temperature is considered given and fixed. (As a matter of interest, this distribution may be compared with the distributions shown in Table 4 for our assembly of seven atoms.)

A solid of molecules with two energy levels each

Since the argument so far has been rather abstract, let us now apply our results to a physically significant example before continuing the more general treatment. In doing so we shall shift our standpoint by specifying the temperature of the system instead of its energy; the justification for assuming that these two specifications are equivalent for large assemblies will be given in Chapter 13.

In this example, which is of practical importance in low temperature physics and which is very instructive, we imagine that each molecule in our assembly (N molecules altogether) can be in either of two non-degenerate energy levels ε_0 and ε_1. (These could, for example, be associated with different magnetic states of the molecules, cf. p. 109 below.) If the assembly has a temperature T we can at once write down the occupation numbers of two levels (omitting the asterisks for simplicity):

$$n_0 = e^{-\alpha} e^{-\varepsilon_0/kT}; \quad n_1 = e^{-\alpha} e^{-\varepsilon_1/kT} \qquad \text{(cf. equation (60))}$$

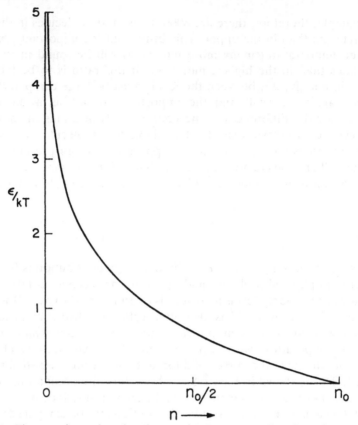

Figure 8 The number of molecules, n, having energy ε, plotted as a function of ε/kT for Boltzmann distribution. This shows how, at a given temperature, n falls off rapidly as ε increases.

Also

$$n_0 + n_1 = N$$

Therefore

$$\frac{n_0}{N} = \frac{e^{-\varepsilon_0/kT}}{e^{-\varepsilon_0/kT} + e^{-\varepsilon_1/kT}} ; \quad \frac{n_1}{N} = \frac{e^{-\varepsilon_1/kT}}{e^{-\varepsilon_0/kT} + e^{-\varepsilon_1/kT}} \qquad (65)$$

As we noted earlier, it doesn't matter what energy zero we choose provided that we are always consistent in its use. To make things as easy as possible, therefore, let us choose our zero to coincide with the level ε_0. This is equivalent to putting $\varepsilon_0 = 0$; to remind ourselves that ε_1 is now measured with respect to ε_0 we will write ε_1 as $\Delta\varepsilon$, the difference between ε_1 and ε_0. Therefore

$$\frac{n_0}{N} = \frac{1}{e^{-\Delta\varepsilon/kT} + 1} ; \quad \frac{n_1}{N} = \frac{e^{-\Delta\varepsilon/kT}}{e^{-\Delta\varepsilon/kT} + 1} = \frac{1}{e^{\Delta\varepsilon/kT} + 1} \qquad (66)$$

These expressions already give us quite a lot of useful information. At high temperatures $(T \to \infty)$, $e^{-\Delta \varepsilon / kT}$ and $e^{+\Delta \varepsilon / kT}$ both tend to unity. Thus $n_0/N \to \frac{1}{2}$ and $n_1/N \to \frac{1}{2}$. This means that at high temperatures the occupation numbers of both states become almost equal $(n_0 = n_1 = N/2)$. By 'high temperatures' we mean $kT \gg \Delta \varepsilon$. This is what we might expect: when the thermal energy available to each molecule is so large that the energy separation between the levels $\Delta \varepsilon$ is negligible in comparison, either level is equally likely to be occupied.

At low temperatures $(T \to 0)$, $e^{-\Delta \varepsilon / kT}$ tends to zero and $e^{+\Delta \varepsilon / kT}$ tends to infinity. Thus $n_0/N \to 1$ and $n_1 \to 0$. This means that, at $T = 0$, all the molecules are in their lowest energy states. (This must be true of any assembly because at $T = 0$ we have simply a mechanical system which is stable only in its lowest energy state.) Figure 9(a) shows diagrammatically how n_1 and n_0 depend on the temperature.

Let us now look at the total energy, E, of the assembly

$$E = n_0 \varepsilon_0 + n_1 \varepsilon_1 = n_1 \Delta \varepsilon \quad \text{(since now } \varepsilon_0 = 0)$$

$$= \frac{N \Delta \varepsilon}{e^{\Delta \varepsilon / kT} + 1} \tag{67}$$

As $T \to 0$, $E \to 0$ as we should expect with all the molecules going into the zero energy state. At high temperature $(kT/\Delta \varepsilon \to \infty)$

$$E \to \frac{N}{2} \Delta \varepsilon$$

Again this is what we should expect because, at high temperatures, half the molecules are in the lower and half in the higher energy state (see Figure 9(b) for the full temperature dependence of E).

From the energy, E, the heat capacity at constant volume can easily be derived:

$$C_V = \left(\frac{\partial E}{\partial T} \right)_V = Nk \left(\frac{\Delta \varepsilon}{kT} \right)^2 \frac{e^{\Delta \varepsilon / kT}}{(e^{\Delta \varepsilon / kT} + 1)^2} \tag{68}$$

For one mole $Nk = R$ so that C_V/R is a function of $(\Delta \varepsilon / kT)$ only. C_V is just the slope of the E versus T curve. Thus it must vanish at high temperatures where E is constant and at low temperatures where both E and its derivatives vanish. C_V has a maximum in the region where $kT \sim \Delta \varepsilon$ (see Figure 9(c) for the full curve). From (68) we can easily show that at low temperatures C_V varies as $(1/T^2) e^{-\Delta \varepsilon / kT}$ which in effect means as $e^{-\Delta \varepsilon / kT}$ because this factor predominates; at high temperatures where $\Delta \varepsilon / kT$ becomes very small, the exponential terms become effectively constant so that C_V falls off as $1/T^2$. This sort of contribution to the heat capacity is found in certain paramagnetic salts at low temperatures. As we shall see in Chapter 15, such salts have been used as a means of producing very low temperatures by the magnetic method.

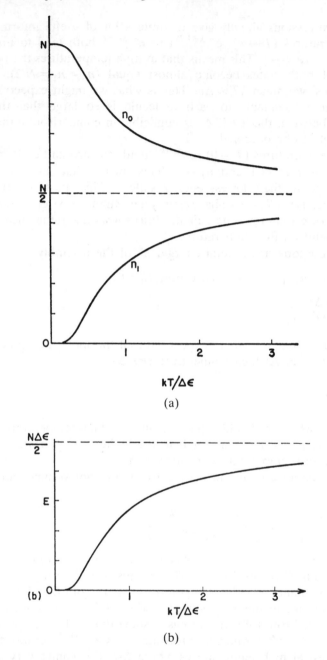

Figure 9 Properties of an assembly of molecules with two energy levels. (a) Occupation numbers n_0 and n_1 as a function of temperature. (b) Energy of the assembly as a function of temperature. (c) Heat capacity as a function of temperature.

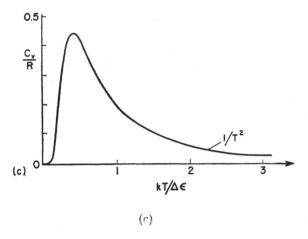

(c)

(c)

Figure 9   (*contd.*)

Notice the following important points about our solid: first, that the heat capacity vanishes at low temperatures. This is because when $kT \ll \Delta\varepsilon$, there is not enough thermal energy to excite many molecules into the upper energy level. Thus, even when the temperature changes, not many molecules can change their energy and this means a small heat capacity. The vanishing of the heat capacity at low temperatures is thus a consequence of the energy gap $\Delta\varepsilon$ between the ground state and the next allowed energy level. Wherever there is such a gap above the ground state the heat capacity will behave in this way even though other features of the model may be quite different from those discussed here.

The second point is that in all our discussion the important ratio has been that of $\Delta\varepsilon$ to kT, i.e. the ratio of two energies, one being the quantum of energy that characterises transitions between the levels, the other the energy that characterises the temperature of the assembly. This is another feature that we shall frequently encounter in other models. In such circumstances it is a convenient trick to represent the molecular energy difference (here $\Delta\varepsilon$) by an equivalent temperature, θ_C say, defined by the relationship:

$$\Delta\varepsilon = k\theta_C$$

where k is Boltzmann's constant. The general thermodynamic properties of the assembly can now be expressed as a function of $kT/\Delta\varepsilon$, i.e. of T/θ_C. The advantage of this device is that the important changes in the temperature dependence of the heat capacity or energy of the assembly can be expected to occur at a temperature such that $kT \sim \Delta\varepsilon$ (this we have already seen), i.e. when $T \sim \theta_C$. So if we know that θ_C is, say, 300 K, we know at once in approximately which temperature range to look for the changes in thermodynamic properties related to the energy difference, $\Delta\varepsilon$.

91

Exercises

Q1 The hydrogen atom has an ionisation energy of 13.6 eV. At what temperature would you expect to find a gas of hydrogen atoms with an appreciable fraction of them ionised by thermal agitation? (Take Boltzmann's constant $k = 1.4 \times 10^{-23}\,\text{J K}^{-1}$ and the charge on the electron as $1.6 \times 10^{-19}\,\text{C}$.)

9

Temperature and entropy in statistical mechanics

Two assemblies that can share energy

So far we have been considering only the occupation numbers of the various energy levels at equilibrium. From this we have been able to deduce information about the thermal energy and heat capacity of a simple model. Now however we will return to some more general properties with a view to interpreting the entropy of an assembly in terms of these distribution numbers. Before considering the entropy, however, let us examine the parameter β (in equations (59) and (60)) more carefully and convince ourselves that it must be closely related to the temperature. To do this, we shall consider briefly what happens if two different assemblies (of localised elements) are allowed to exchange energy with each other. Let assembly A have N molecules, each of which may have energy ε_1, ε_2, ε_3 etc., and let assembly B have M molecules, each of which may have energies η_1, η_2, η_3 etc. Consider now a typical distribution in which in assembly A

n_1 molecules have energy ε_1
n_2 molecules have energy ε_2
. . .
n_i molecules have energy ε_i
. . .

and in assembly B

m_1 molecules have energy η_1
m_2 molecules have energy η_2
. . .
m_j molecules have energy η_j
. . .

We now have three restrictive conditions:

$$\sum_i n_i = N; \tag{69}$$

$$\sum_j m_j = M; \tag{70}$$

and

$$\sum_i n_i \varepsilon_i + \sum_j m_j \eta_j = E \tag{71}$$

The first two conditions mean simply that the number of A molecules and the number of B molecules are separately fixed. The third condition is most important. This expresses the fact that the two assemblies can share energy and the total for the two together is fixed and equal to E.

We now write down the number of stationary states of the two assemblies combined; this number is given by

$$P = \frac{N!}{n_1! n_2! \ldots n_i! \ldots} \times \frac{M!}{m_1! m_2! \ldots m_j! \ldots} \tag{72}$$

This is just the product of the number of stationary states in assembly A with the number of stationary states of assembly B. It means that any stationary state of one can be combined with any stationary state of the other to produce a stationary state of the composite assembly.

We now want to find the values of the n_i's and of the m_j's which make P a maximum subject to the restrictive conditions.

To do this we proceed as before. We take $\ln P$ and allow all the n_i's and m_j's to vary by small amounts.

Thus

$$\delta \ln P = - \sum_i \delta \ln n_i! - \sum \delta \ln m_j! = 0 \tag{73}$$

subject to the condition that

$$\sum \delta n_i = 0 \tag{74}$$

$$\sum \delta m_j = 0 \tag{75}$$

$$\sum \varepsilon_i \delta n_i + \sum \eta_j \delta m_j = 0 \tag{76}$$

Using Stirling's approximation, equation (73) can be rewritten

$$\sum \ln n_i \, \delta n_i + \sum \ln m_j \, \delta m_j = 0 \tag{77}$$

To take account of the restrictive conditions we multiply (74) by α_A, (75) by α_B and (76) by β. Add these to (77) and we get

$$\sum (\ln n_i + \alpha_A + \beta\varepsilon_i)\, \delta n_i + \sum (\ln m_j + \alpha_B + \beta\eta_j)\, \delta m_j = 0 \qquad (78)$$

Proceeding as before, we choose α_A, α_B and β so that the first three brackets vanish. The remaining δn_i's and δm_j's are then independent. If the sum is to be zero for arbitrary values of the δn_i's and the δm_j's we must then have

$$\ln n_i = \alpha_A + \beta\varepsilon_i = 0; \quad \ln m_j + \alpha_B + \beta\eta_j = 0$$
$$n_i^* = e^{-\alpha_A - \beta\varepsilon_i} \qquad m_j^* = e^{-\alpha_B - \beta\eta_j} \qquad (79)$$

The striking feature of this result is that the parameter β is common to both assemblies. This arises because the two assemblies can share energy. If they had had separate total energies E_A and E_B we should have had to introduce separate restrictive conditions

$$\sum n_i\varepsilon_i = E_A, \quad \sum m_j\eta_j = E_B$$

with two corresponding Lagrange multipliers β_A and β_B.

We can therefore conclude that if two or more assemblies are allowed to share energy the distribution numbers in each are characterised by a common value of β. The sharing of energy between the assemblies is analogous to bringing two thermodynamic systems into thermal contact; at equilibrium, such systems, as we know, have the same temperature. Thermal equilibrium thus implies equal temperatures and in our model equal values of β. It is therefore reasonable to regard β as being a sort of empirical temperature. To go further we want to find out how β is related to the *absolute temperature*; for this we have to invoke the second law of thermodynamics, which we will now do.

Entropy and absolute temperature

Let us return now to our single assembly of localised elements. Their number is N, their total energy E, and the volume accessible to them is V. We know that for a given set of n_i's the corresponding number of stationary states of the assembly is

$$P = \frac{N!}{n_1! n_2! \ldots n_i! \ldots}$$

As we have already shown, the values of the distribution numbers which make P a maximum subject to the restrictive conditions are given by

$$n_i^* = e^{-\alpha - \beta\varepsilon_i}$$

Now we are going to consider what happens if we allow the energy, E, of the assembly to change by a small amount, δE, and the volume by a small amount, δV. In the process the assembly is to be always effectively in equilibrium; it is thus, in thermodynamic terms, a quasi-static (reversible) change. The thermodynamic description of the process is thus

$$\delta E = T\,\delta S - p\,\delta V \qquad (80)$$

where T is the absolute temperature of the assembly, p the pressure and δS the entropy change that occurs.

Let us now compare this with the description of the same change given by *statistical mechanics* and see how the two can be reconciled.

As we change the energy of the assembly, the n_i's will change and so will the energy levels ε_i. Throughout the change

$$E = \sum_i n_i^* \varepsilon_i$$

– the equilibrium distribution numbers come in because the change is quasi-static – so that

$$\delta E = \sum_i \varepsilon_i\,\delta n_i^* + \sum_i n_i^*\,\delta\varepsilon_i \qquad (81)$$

We shall now manipulate this equation to put it into a form more directly comparable with equation (80). It will turn out that the first sum on the right-hand side of (81) corresponds to the heat entering the system ($T\,\delta S$ in (80) and the second sum to the work done ($-p\,\delta V$ in (80)).

Interpretation of $\sum n_i^* \delta\varepsilon_i$

Let us first deal with the sum $\sum n_i^* \delta\varepsilon_i$. We have assumed throughout that the only way that the ε_i's can change is through a change of volume. Therefore we can write

$$\delta\varepsilon_i = \frac{\mathrm{d}\varepsilon_i}{\mathrm{d}V}\,\delta V$$

and so

$$\sum_i n_i\,\delta\varepsilon_i = \left\{ \sum n_i \frac{\mathrm{d}\varepsilon_i}{\mathrm{d}V} \right\} \delta V \qquad (82)$$

Let us now introduce the volume per molecule $v = V/N$ and rewrite equation (82) as

$$\sum_i n_i^*\,\delta\varepsilon_i = \left\{ \sum_i \frac{n_i^*}{N} \frac{\mathrm{d}\varepsilon_i}{\mathrm{d}(V/N)} \right\} \delta V = \left\{ \sum_i \frac{n_i^*}{N} \frac{\mathrm{d}\varepsilon_i}{\mathrm{d}v} \right\} \delta V \qquad (83)$$

We now interpret $-(d\varepsilon_i/dv)$ as the effective pressure, p_i, exerted by each of the n_i molecules in the ith energy level. This cannot be demonstrated directly from the expressions for the energy levels of the *solids* which we shall deal with because the volume dependence of these energy levels has to be introduced artificially. However, this result can be checked directly from the expression for the energy levels of an *ideal gas* (equation (140) below).[5] On this basis, $\Sigma n_i p_i/N$ is just the average pressure, p, of the whole assembly. We thus identify the second sum in (81) as $-p\,\delta V$, the work done during the quasi-static process. (If you find this identification unconvincing or unsatisfying you may take it at this stage as a postulate. This postulate is consistent with the interpretation we are now about to give of the other sum $(\Sigma \varepsilon_i \delta n_i)$ and the complete interpretation of equation (81) can then be tested by the validity or otherwise of its consequences.)

Interpretation of $\Sigma \varepsilon_i \delta n_i^*$

Now turn to the first sum on the right-hand side of (81). Because the process is quasi-static, i.e. a succession of equilibrium states, then at each stage

$$n_i^* = e^{-\alpha - \beta \varepsilon_i} \tag{84}$$

We can thus express the ε_i's in terms of the n_i^*'s and β:

$$\ln n_i^* = -\alpha - \beta_i \varepsilon_i \qquad \text{(from equation (84))}$$

so that

$$\varepsilon_i = -\frac{(\ln n_i^* + \alpha)}{\beta} \tag{85}$$

α and β will, of course, now be changing during the process. Substituting in the sum we find

$$\sum \varepsilon_i \, \delta n_i^* = -\frac{1}{\beta} \sum (\ln n_i^* + \alpha) \, \delta n_i^*$$

$$= -\frac{1}{\beta} \left\{ \sum \ln n_i^* \, \delta n_i^* + \alpha \sum \delta n_i^* \right\}$$

The second sum in the bracket vanishes because $N = \Sigma n_i^*$ is a constant throughout. Finally therefore

$$\sum \varepsilon_i \, \delta n_i^* = -\frac{1}{\beta} \left\{ \sum \ln n_i^* \, \delta n_i^* \right\} \tag{86}$$

However, we can simplify this a little more by relating the factor in brackets to the value of P. For a given set of distribution numbers, the number of stationary states P is given by

$$\ln P = \ln N! - \sum \ln n_i!$$

To obtain the maximum value of P for a given energy and volume we must put in the n_i^*'s:

$$\ln P_{max} = \ln N! - \sum \ln n_i^*!$$

If E and V are kept constant but the n_i's are allowed to change, $\delta \ln P_{max}$ must vanish (because this is how we found the n_i^*'s). If, however, E and V are *not* kept constant, $\delta \ln P_{max}$ no longer vanishes. In a quasi-static change under these conditions

$$\delta(\ln P_{max}) = -\sum \ln n_i^* \, \delta n_i^* \tag{87}$$

since N is fixed. (In this result we have differentiated $\ln n_i^*!$ using Stirling's approximation.) If we look back to equation (86) we see that we can now write

$$\sum \varepsilon_i \, \delta n_i^* = \frac{1}{\beta} \, \delta(\ln P_{max}) \tag{88}$$

This completes the transformation of equation (81) so now let us compare it with the thermodynamic description of the same change, namely equation (80) above.

Equation (81) has now become

$$\delta E = \frac{1}{\beta} \, \delta(\ln P_{max}) + \left\{ \sum \frac{n_i}{N} \frac{d\varepsilon_i}{dv} \right\} \delta V \tag{81A}$$

and equation (80) is

$$\delta E = T \, \delta S - p \, \delta V \tag{80}$$

We have already seen that

$$\sum \frac{n_i}{N} \frac{d\varepsilon_i}{dv} = -p$$

so that the two equations (81A and 80) can be made identical if we make $\delta(\ln P_{max})$ proportional to δS and β proportional to $1/T$.

More specifically, we can write

$$\left. \begin{array}{l} \delta S = k \, \delta(\ln P_{max}) \\[1em] \text{and} \\[0.5em] \beta = 1/kT \end{array} \right\} \tag{89}$$

where k is a constant independent of E, S, V or T, or of the type of assembly. Notice that k enters into both results so that it vanishes in the product $T \, \delta S$.

The constant k is known as Boltzmann's constant and is of central importance in evaluating entropies by statistical mechanics. As we shall see when we deal with an ideal gas, $N_0 k = R$, where R is the gas constant per mole and N_0 is Avogadro's number. k thus turns out to have the value

$1.38 \times 10^{-23} \, \text{J K}^{-1}$

We have already seen how the quantity kT is an important measure of the energy associated with thermal effects. Now we see that k is also the parameter which determines the entropy scale.

We have now specified β in terms of the *absolute* temperature T and simultaneously found a 'mechanical' interpretation of the entropy S. To complete the operation we integrate the expression for δS (see equation (89)) and get

$S - S_0 = k \ln P_{\text{max}}$

As we shall see below, the third law of thermodynamics requires that the entropy of all states of a system in internal thermodynamic equilibrium should be the same at the absolute zero of temperature. This as we shall see carries the implication that S_0 in the above equation must be the same for all assemblies; for convenience therefore we set it equal to zero. Our fundamental equation thus becomes

$S = k \ln P_{\text{max}}$ (90)

This equation provides the basis for our statistical interpretation of entropy.

Boltzmann and Planck referred to the quantity P as the 'probability' or 'thermodynamic probability' of the corresponding state of the system. The essence of the second law of thermodynamics is thus that thermodynamic systems of fixed energy always tend to move from less probable to more probable states and the equilibrium state corresponds to that of maximum probability. Boltzmann indeed arrived at his interpretation of entropy by considering how gases behaved when *not* in equilibrium and how they approached equilibrium. His work led to a generalisation of the concept of entropy because he was able to define the entropy of a gas even when it was not at equilibrium.

In what follows however we shall be concerned almost entirely with the entropy of systems in equilibrium. Under these conditions the number of microstates corresponding to the equilibrium configuration is so large compared to the number corresponding to any appreciable departure from equilibrium that we can use an alternative (and sometimes more convenient) form of equation (90). Instead of

$S = k \ln P_{\text{max}}$

we can just as accurately write

$S = k \ln W$ (90A)

where now W is the *total* number of microstates of the assembly which are consistent with the given values of E, V and N, i.e. $W = \Sigma P$. The difference between (90) and (90A) is that W is equal to the sum of all possible values of P, not just the maximum value. For large-scale assemblies, however, $\ln P_{max}$ and $\ln W$ are, to all intents and purposes, identical and we can use whichever expression is more convenient.[6] Note that it is the *logarithms* that are almost identical, not of course the numbers themselves.

A different point of view

This is the heart of our statistical treatment of entropy. Our discussions have shown that there should be an intimate relationship between the *entropy* of a thermodynamic system (of given E, V and N) and the *number of microscopic states*, W, that are compatible with the prescribed values of E, V and N. On the basis of certain assumptions we have shown how this relationship can be deduced for a certain type of assembly (of localised, independent particles). We shall henceforth make use of this relationship as if it were quite generally valid and deduce from it the entropy of some fairly simple models of solids and gases. The validity of the relationship is to be thought of as established by the success of its predictions.

However, before using the relationship $S = k \ln W$ let us look briefly at a treatment, due to Planck (1858–1947), which helps, perhaps, to make clear why the relationship must have this particular form. Suppose that we are convinced that S must be related to W. In fact let us assume (and this is a big assumption) that

$$S = f(W) \tag{91}$$

where, in Planck's words, '$f(W)$ represents a universal function of the argument W'.[7] We now apply this relationship to two nearly independent systems whose entropies are S_1 and S_2 with corresponding values of W, W_1 and W_2. The two systems could, for example, be two blocks of copper side by side and touching each other. The entropy of the two taken together is obviously

$$S_{12} = S_1 + S_2 \tag{92}$$

On the other hand the number of microstates for the composite system is just

$$W_{12} = W_1 W_2 \tag{93}$$

because any given microstate of system 1 may be combined with each microstate in turn of system 2 to give a distinguishable microstate of the composite assembly.

To reconcile equations (91), (92) and (93) we require that

$$f(W_1 W_2) = f(W_1) + f(W_2)$$

The general solution of this equation is

$$f(W) = k' \ln W$$

where k' is an arbitrary constant. Consequently S and W must on the basis of our original assumption be related in the following way:

$$S = k' \ln W$$

Entropy and disorder

We have just seen from our statistical discussion of entropy that the entropy of an assembly is related to the number of microstates of the assembly which are compatible with the fixed values of energy, volume and number of particles. Although the relationship has been demonstrated only for solid-like assemblies, it is in fact valid for gas-like assemblies too. In order to get an intuitive way of thinking about entropy this relationship has been interpreted in various ways. Some people think of entropy as being measured by the 'disorder' of the system; that high entropy means great disorder, and a system with low entropy means an ordered system. Guggenheim[8] has suggested that entropy is to be thought of as 'spread'; for example, high entropy means that the elements of an assembly are spread over a wide range of energy levels. Both of these ideas are useful and obviously it is a matter of personal preference which is more helpful. When we come to the third law of thermodynamics and consider what happens to the entropy at low temperatures the concept of ordering is probably more useful. Moreover the tendency for entropy to increase in spontaneous changes can then be interpreted as meaning that in general things become disordered more readily than they become ordered, simply because there are very many more ways of producing disorder than of producing an ordered arrangement. It also explains why in general it is easier to produce high temperatures than to produce low temperatures.

Let us apply this idea of disorder or 'spread' to a simple and familiar example. Consider how the entropy of a given quantity of steam alters as it becomes cooler and cooler and changes first into water and then into ice. Suppose that the steam is at atmospheric pressure above 100°C; we now allow it to cool at constant pressure. At 100°C it will condense into water and its volume will contract nearly two thousand-fold. At the same time its entropy will be reduced and the entropy change will show up in the latent heat of condensation.

In thinking of the entropy of the system in statistical terms we must consider both the spread of the molecules in actual space (i.e. the volume accessible to them) and also the spread of their velocities. The spread of velocities of the molecules in the steam at 100°C is the same as that in the water at the same temperature, so that the entropy associated with this spread is the same. On the other hand, the entropy associated with the spatial configuration diminishes greatly because of the volume change and so we see that the entropy of condensation has its counterpart in the 'ordering' process represented by this volume change.

On cooling the liquid still further, its volume does not change greatly but the velocity spread diminishes as the temperature goes down (cf. Figure 7) so that the entropy continues to fall slowly. At 0°C the water freezes. At this temperature the ordering process is more obvious because when the entropy of melting has been removed the system is now a crystalline solid in which the atoms are arranged in ordered arrays on a lattice. On cooling still further, the vibrations of the atoms about their lattice sites are reduced in amplitude and other ordering processes associated with the rotational states of the molecules, with their nuclear spins etc., will occur if the disorder is not 'frozen in' before thermodynamic equilibrium is attained.

Thus we see that the fall in entropy of the substance when it is cooled corresponds on the microscopic scale to a progressive ordering of the molecules in the substance until at very low temperatures we find a highly ordered arrangement of atoms and molecules in the crystalline solid. These questions will be further discussed when we come to the third law of thermodynamics.

With these general ideas in mind let us now go on to examine in detail the entropy and other properties of a number of different assemblies. We shall start with solids because we have learned how to enumerate the microstates in them. In gases the counting is slightly different and we shall consider them separately.

Exercises

Q1 The proton has spin 1/2 and so in a magnetic field it may lie parallel or anti-parallel to the field. In a nuclear resonance experiment at 300 K, a magnetic field of 1 T is applied to a solid containing hydrogen. Treat the protons in this as non-interacting and so calculate the fractional difference in occupation numbers of the two magnetic energy levels of the protons, $k = 1.36 \times 10^{-23}$ J K^{-1} and the magnetic moment of the proton is 1.41×10^{-26} J T^{-1}. Why is this difference important in nuclear magnetic resonance?

Application to solids

The entropy of a solid-like assembly

Let us now deduce the entropy of a solid composed of N identical particles each having energy levels ε_0, ε_1, ..., ε_i, ... We use the same notation as in the section on entropy and absolute temperature.

Then the entropy S is given by

$$S = k \ln P_{\max} \tag{94}$$

where

$$P_{\max} = N! / \prod n_i^* \tag{95}$$

So

$$S = k \ln N! - k \sum \ln n_i^*!$$

If we make use of Stirling's formula, we find

$$\frac{S}{k} = N \ln N - \sum n_i^* \ln n_i^* \tag{96}$$

But

$$n_i^* = e^{-\alpha - \beta \varepsilon_i} \quad \text{or} \quad \ln n_i^* = -\alpha - \beta \varepsilon_i$$

where $\beta = 1/kT$.

Therefore

$$\frac{S}{k} = N \ln N + \alpha \sum n_i^* + \frac{1}{kT} \sum n_i^* \varepsilon_i \tag{97}$$

We know that

$$\sum n_i^* = N \quad \text{and} \quad \sum n_i^* \varepsilon_i = E$$

so that

$$\frac{S}{k} = N \ln N + \alpha N + \frac{E}{kT} \tag{98}$$

To get rid of α we use the result that

$$N = \sum n_i^* = e^{-\alpha} \sum e^{-\varepsilon_i/kT}$$

giving

$$\alpha = -\ln N + \ln \sum e^{-\varepsilon_i/kT} \tag{99}$$

Thus

$$\frac{S}{k} = N \ln \sum e^{-\varepsilon_i/kT} + \frac{E}{kT} \tag{100}$$

This is the result for the entropy of our solid; it can be rewritten in an equivalent form by making use of the thermodynamic identity that defines the Helmholtz free energy:

$$F = E - TS$$

Thus rearranging (100) we get

$$F = -NkT \ln \sum e^{-\varepsilon_i/kT} \tag{101}$$

This is probably the simplest way to deduce the entropy of an assembly for which we know the values of the ε_i's. We work out F from this expression and then use the thermodynamic identity $S = -(\partial F/\partial T)_V$ to find the entropy. It is not difficult to verify directly that if we evaluate $-(\partial F/\partial T)_V$ from (101) we do in fact get back to (100).

Notice that the result for F depends only on the quantity

$$Z = \sum e^{-\varepsilon_i/kT} \tag{102}$$

This is often called the 'partition function' because each term is proportional to the value of the occupation number n_i for that level; thus the particles are 'partitioned' among the energy levels in proportion to the terms in the sum. In terms of the partition function Z,

$$F = -NkT \ln Z \tag{103}$$

and

$$S = -\left(\frac{\partial F}{\partial T}\right)_V = Nk\frac{\partial}{\partial T}(T\ln Z) \tag{104}$$

Also it can be shown quite readily that since $E = \Sigma n_i\varepsilon_i$

$$E = NkT^2\frac{\partial}{\partial T}(\ln Z) \tag{105}$$

So we see that once we know Z, which can be found from the values of the energy levels, we can then deduce all the thermodynamic properties of our assembly.

So far we have assumed that for each molecule there is only one stationary state corresponding to each energy level. If, however, the energy levels are degenerate so that to the ith level of energy ε_i there correspond ω_i stationary states, the partition function then becomes

$$Z = \sum\omega_i\mathrm{e}^{-\varepsilon_i/kT} \tag{102A}$$

Equations (103), (104) and (105) can now be used with this more general partition function.

The Einstein solid

To illustrate how we can use the ideas of statistical mechanics that we have already developed, let us use them to calculate the entropy and heat capacity of a simple model of a solid. This model of a solid was introduced by Einstein (1879–1955) in a paper of 1907.[9] He writes: 'The simplest idea that one can form of the thermal motion in solid bodies is that the atoms thereof execute harmonic oscillations about their equilibrium positions.' Because the atoms can move in three independent directions in space, each atom can be regarded as equivalent to three independent one-dimensional harmonic oscillators. If we have M identical atoms in the solid, then as far as their thermal motion is concerned, we picture them as behaving like $3M$ simple harmonic oscillators. If their classical frequency of vibration is ν the energy levels available to each are given (according to quantum mechanics) by

$$\varepsilon_i = (i + \tfrac{1}{2})h\nu \tag{106}$$

where $i = 0, 1, 2, 3, \ldots$. There is only one stationary state for each energy level (to each value of i).

We now wish to evaluate the partition function for our assembly (equation (102)). In this example it is

$$Z = \sum\mathrm{e}^{-\varepsilon_i/kT} = \mathrm{e}^{-h\nu/2kT} + \mathrm{e}^{-3h\nu/2kT} + \ldots + \mathrm{e}^{-(2i+1)h\nu/2kT} + \ldots$$

105

This therefore forms a convergent geometric progression whose sum is thus:

$$Z = \frac{e^{-h\nu/2kT}}{1 - e^{-h\nu/kT}}$$

Notice that Z depends only on $h\nu/kT$. Let us therefore define a characteristic temperature, θ_E (E for Einstein solid), such that $h\nu = k\theta_E$. We can then write

$$Z = \frac{e^{-\theta_E/2T}}{1 - e^{-\theta_E/T}} \tag{107}$$

Consequently the Helmholtz free energy of the $3M$ linear oscillators which represent the solid is given by

$$F = -3MkT \ln Z = 3MkT \ln(1 - e^{-\theta_E/T}) + \frac{3Mk\theta_E}{2} \tag{108}$$

Thus we have deduced the free energy of our solid; the second term in F which is independent of temperature and gives the value of F at 0 K is called the zero-point energy. It arises because when all the oscillators are in their lowest energy states ($i = 0$ in equation (106)) there is still a 'zero-point' energy of $\frac{1}{2}h\nu$ per oscillator.

Now we use $S = -(\partial F/\partial T)_V$ to find the entropy and we get

$$S = 3MkT \frac{\theta_E}{T^2} \frac{e^{-\theta_E/T}}{1 - e^{-\theta_E/T}} - 3Mk \ln(1 - e^{-\theta_E/T})$$

or

$$\frac{S}{3Mk} = \frac{\theta_E}{T}(e^{\theta_E/T} - 1)^{-1} - \ln(1 - e^{-\theta_E/T}) \tag{109}$$

At T tends to zero this expression goes to zero; this is an example of the third law of thermodynamics which we shall discuss more fully later. Note that the zero-point energy does *not* contribute to the entropy. At high temperatures $(T \gg \theta_E)$ $S \approx 3Mk \ln(T/\theta_E)$. The full dependence of the entropy on temperature is illustrated in Figure 10(a).

The internal energy of the assembly can also be found, since, from equation (105),

$$E = 3MkT^2 \frac{\partial}{\partial T}(\ln Z)$$

$$= \frac{3Mk\theta_E}{2} + \frac{3Mk\theta_E}{e^{\theta_E/T} - 1} \tag{110}$$

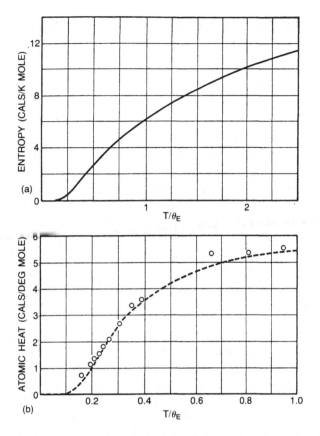

Figure 10 (a) The entropy of an Einstein solid as a function of reduced temperature T/θ_E. (b) The atomic heat capacity of diamond as a function of reduced temperature T/θ_E. The dashed line is the theoretical curve; the points give the experimental values. (Copy of Einstein's original figure in *Annalen der Physik*, **22**, 180, by permission of the Editor.)

The first term here is just the zero-point energy again; the second term, which depends on T, may be called the thermal energy.

The heat capacity is then given by

$$C_V = \left(\frac{\partial E}{\partial T}\right)_V$$

If we have one gram-atom of solid, M is Avogadro's number and so $Mk = R$, the gas constant. Thus by differentiating equation (110) we find

$$C_V = 3R\left(\frac{\theta_E}{T}\right)^2 \frac{e^{\theta_E/T}}{(e^{\theta_E/T} - 1)^2} \tag{111}$$

At high temperatures $(T \gg \theta_E) C_V \rightarrow 3R$ and becomes constant. At low temperatures

$$C_V = 3R \left(\frac{\theta_E}{T} \right)^2 e^{-\theta_E/T}$$

which tends to zero as $T \rightarrow 0$.

θ_E was defined by the relation $\theta_E = h\nu/k$ where ν was the characteristic vibrational frequency of an atom in the solid. This frequency depends on the mass of the atom and on the restoring forces acting on it. Since both these vary from solid to solid, θ_E must also vary from solid to solid. Each simple solid then may be thought of as having a characteristic value of θ_E.

If you look at equation (111) you will see that apart from the gas constant R, C_V depends only on θ_E/T. If therefore we plot C_V as a function of T/θ_E we have a single curve which is valid for all temperatures and solids. To find the heat capacity for a particular solid at a particular temperature all we need to know is its characteristic temperature θ_E and we can then deduce C_V from this universal curve.

Figure 10(b) shows the curve as it was originally illustrated in Einstein's paper. For comparison with the experiment Einstein also shows the experimental values of the heat capacity of diamond (1 gram-atom). In making the comparison between theory and experiment he has chosen the value of θ_E for diamond to make the curves agree at one temperature. It is clear from the figure and equation (111) that this simple Einstein model reproduces the main features of the temperature dependence of C_V as found by experiment. In particular, at high temperatures C_V becomes independent of T and approaches the value of $3R$ as can readily be deduced from equation (111). This corresponds to the law of Dulong and Petit. At low temperatures on the other hand the heat capacity falls off to zero. Notice too that the temperature at which the fall-off begins is roughly where $T/\theta \sim 1$. This means at a temperature T such that kT is about equal to $h\nu$; that is $kT \sim \varepsilon$, where ε is the energy separation between the energy levels of the individual atomic oscillators.

This model due to Einstein gave the first explanation of the fall-off in the values of the specific heat of solids at low temperatures. It is, of course, an oversimplification to regard each atom in the solid as an independent oscillator; since Einstein's paper was published the model has been refined to take account of the coupling of the motion of one atom with the other atoms in the lattice. (One result from these more refined models that will be useful to us later is that at low enough temperatures the lattice heat capacity of all solids must vary as T^3.) Nevertheless the Einstein model describes correctly many of the essential features of the heat capacity of a simple solid.

A simple magnetic solid

The thermal vibrations of the atoms and molecules in solids provide one of the most important contributions to the entropy of these solids. We have just considered a simple model of a solid whose molecules could vibrate about their positions of equilibrium (an Einstein solid), and have found out how the entropy of this solid varies with temperature. Another important contribution to the entropy of solids occurs in those solids whose molecules possess a magnetic moment. Once again we shall consider the simplest example. In this model we think of each molecule of the solid as being effectively a small dipole with magnetic moment μ. In many paramagnetic solids the molecules do indeed act as dipoles although in general it is just one ion in the molecule or aggregate of molecules which carries the magnetic moment. For example in iron ammonium alum, $FeNH_4(SO_4)_2 . 12H_2O$, it is the Fe ion that carries the magnetic moment.

Consider now the application of a magnetic field, B, to the solid. In this field each molecular dipole has energy $-\mu B \cos \theta$, where θ is the angle between B and the direction of μ. When μ and B are parallel ($\theta = 0$) the energy is $-\mu B$ (its minimum value). When μ and B are anti-parallel ($\theta = \pi$) the energy is $+\mu B$ (its maximum value). In classical mechanics all energies (corresponding to all values of θ) between these extremes would be allowed. In quantum mechanics, characteristically, only certain discrete energies (or directions) are allowed. Take the simplest case: in this each molecule has only *two* possible orientations in the presence of a magnetic field; these correspond to having μ parallel and anti-parallel to the field direction. (This would occur if the ion had a resultant spin of 1/2.)

Suppose that a uniform magnetic field B is applied to the solid. Some of the molecular magnets will point along the field and some in the opposite direction. The energy of the first group ($-\mu B$) is lower than that of the second group ($+\mu B$). At 0 K (in the absence, that is, of any thermal energy) all the dipoles would thus point along the field since this would put all the molecules into their lowest magnetic energy states. At higher temperatures, the thermal energy perturbs this state of affairs and we must now appeal to statistical mechanics to find out how they are distributed.

First, let us consider what happens in a vanishingly small field. As B tends to zero, the energy difference between the levels will tend to zero and so in the absence of a magnetic field the two magnetic energy levels coalesce (the level becomes twofold degenerate). What is the magnetic entropy of the assembly of N molecules in this condition? To answer this we need to know W, the total number of stationary states of the assembly corresponding to the conditions of the problem. Each molecule can occupy either of two states of equal energy. The first molecule can thus have two possible stationary states, the second molecule two, and so on. Each of these possibilities can be combined with each of the others so that the total number for the assembly is

$$W = 2^N$$

Now use $S = k \ln W$ since this happens to be more convenient here than $S = k \ln P_{max}$. Therefore

$$S_{magnetic} = k \ln 2^N$$

$$= Nk \ln 2 \tag{112}$$

If N is Avogadro's number N_0

$$S_{magnetic} = R \ln 2 \tag{113}$$

It is clear from this that if the state had been 3, 4, 5, . . ., n-fold degenerate instead of 2-fold, we should have found

$$S_{magnetic} = R \ln 3, \; R \ln 4, \; R \ln 5 \text{ etc.} \tag{114}$$

Now let us write down the equilibrium occupation numbers when a field B is applied to the solid at temperature T. The energy levels available to each molecule are

$$\varepsilon_0 = - \mu B, \quad \varepsilon_1 = + \mu B$$

We can therefore immediately write down the occupation numbers of the two levels:

$$\frac{n_0^*}{N} = \frac{e^{\mu B/kT}}{e^{\mu B/kT} + e^{-\mu B/kT}}; \quad \frac{n_1^*}{N} = \frac{e^{-\mu B/kT}}{e^{\mu B/kT} + e^{-\mu B/kT}} \tag{115}$$

or

$$\frac{n_0^*}{N} = \frac{1}{e^{-2\mu B/kT} + 1}; \quad \frac{n_1^*}{N} = \frac{1}{e^{2\mu B/kT} + 1} \tag{116}$$

It is clear that we are now dealing with a solid whose molecules can take up either of two non-degenerate energy states separated by an energy $2\mu B$. We can thus take over all the expressions already derived for the assembly of molecules each having two non-degenerate energy levels. (See equations (19) and (20).) In our present example the energy separation $\Delta \varepsilon$ must be put equal to $2\mu B$.

Clearly the temperature dependence of the entropy, S, the energy, E, and the heat capacity, C_V, will now depend on the magnitude of the applied field, although the limiting value of S at high temperatures is $R \ln 2$ for all finite fields. As $T \to 0$, there will always be temperatures such that $kT \ll 2\mu B$ for all non-vanishing values of B. Under these conditions, effectively all the molecules are in their lowest level. Each molecule can be put into its ground state in only one way so that for the whole assembly there is only one possible microscopic state. Thus as $T \to 0$, $W \to 1$, $S = k \ln W$ tends to zero. Thus provided there is a non-vanishing field $S \to 0$ as $T \to 0$. If our model were strictly correct, the two energy levels would coalesce for $B = 0$ and then the entropy would remain at $R \ln 2$ right down to absolute zero. This, as we shall see, would contradict the third law of thermodynamics. In actual solids, of course, the interaction between the molecular dipoles would make the two directions of μ in the solid have different energies. As a first approximation we can in some cases treat this interaction as due to an internal field B_0 which thus, even when no external field is applied, separates the two energy states. This then makes $S \to 0$ as $T \to 0$ under all circumstances, as is required by the third law of thermodynamics (see Chapter 14).

Let us now work out the entropy in more detail. In general, the number of microstates corresponding to a given E, B and N is in this case given by

$$P_{\max} = \frac{N!}{n_0^*! n_1^*!}$$

where n_0 and n_1 are given by equation (116). We will now drop the asterisks and remember that the n's are equilibrium distribution numbers. Therefore

$$\frac{S}{k} = \ln P_{\max} = N \ln N - n_0 \ln n_0 - n_1 \ln n_1 \qquad \text{(using Stirling's formula)}$$

$$= (n_0 + n_1) \ln N - n_0 \ln n_0 - n_1 \ln n_1$$

$$= - n_0 \ln (n_0/N) - n_1 \ln (n_1/N) \qquad (117)$$

(Notice that this is entirely symmetrical in n_0 and n_1.) Putting in values of n_0 and n_1 (from equation (116)), we have

$$S = \frac{Nk}{e^{-2\mu B/kT} + 1} \ln (e^{-2\mu B/kT} + 1) + \frac{Nk}{e^{2\mu B/kT} + 1} \ln (e^{2\mu B/kT} + 1) \qquad (118)$$

At low temperatures ($kT \ll 2\mu B$), the first term becomes negligible and

$$S \simeq Nk \left(\frac{2\mu B}{kT} \right) e^{-2\mu B/kT} \qquad (119)$$

At high temperatures ($kT \gg 2\mu B$):

$$S \to Nk \ln 2 \qquad (120)$$

Points to notice about the entropy versus temperature curve are:

(i) S goes to zero as T goes to zero for all non-vanishing values of B as already discussed.

(ii)   The total chanage in S (i.e. from 0 to $R \ln 2$) does not depend on temperature but just on the number of levels involved (this is always so for a finite number of levels).

(iii)   When S begins to change, C_B (which is equal to $T(\partial S/\partial T)$) becomes appreciable: where S changes rapidly, C_B becomes large.

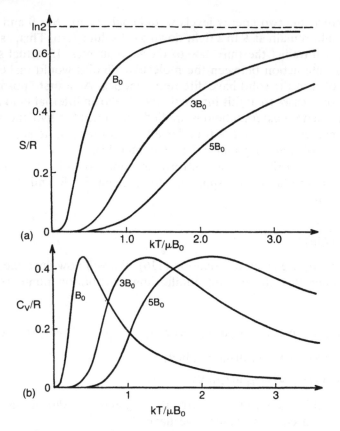

Figure 11 (a) The entropy of a simple magnetic solid as a function of temperature for three values of the applied external field (B_0, $3B_0$ and $5B_0$). (b) The corresponding heat capacities.

Thus, in this as in other similar examples, a sudden fall in S with falling temperature means a heat capacity 'anomaly'. Figures 11(a) and (b) show how S and C_B vary with temperature at three different values of B.

Exercises

Q1 An insulating solid consists of identical atoms whose allowed energy states are a single non-degenerate level at ε_1 and a doubly degenerate level at a higher energy ε_2. Write down the partition function for the solid. What is the limiting value of the entropy of 1 mole of the solid at high temperatures and at low temperatures? What do *high* and *low* mean here?

Q2 An inert gas solid is found to have the following properties. The characteristic temperature of its lattice is approximately 80 K and its nuclei have spin 1/2. The nuclear spin energy levels in zero magnetic field are separated by an energy of 10^{-11} eV. Estimate approximately for 1 mole: (a) the entropy and heat capacity at 1 K; (b) the heat capacity at 300 K. At about what temperature would you expect to find the nuclear spin entropy changing rapidly?

$$k = 1.4 \times 10^{-23} \, \text{JK}^{-1} \quad \text{and} \quad e = 1.6 \times 10^{-19} \, \text{C}$$

Q3 A solid has atoms which have access to only three non-degenerate energy levels, ε_0, ε_1 and ε_2. Show that, as T tends to infinity, the occupation numbers $n_0 = n_1 = n_2$ and that the average energy per particle is then $E_{\text{lim}} = (\varepsilon_0 + \varepsilon_1 + \varepsilon_2)/3$. Imagine that the solid is now at some finite temperature and is thermally isolated. Work is done on the solid (by a magnetic field, for example) and the energy per particle is made to exceed E_{lim}. Show that then either n_1 or n_2 or both must exceed n_0. In fact the distribution of particles follows a positive exponential dependence on energy and can be formally described by a negative temperature. The better name is a *transfinite* temperature. Such a system can be realised experimentally at low temperatures with a solid whose paramagnetic ions can provide a magnetic system essentially isolated from the lattice vibrations and on which magnetic work can be done by an external magnetic field.

11

Application to gases: (1) the classical approximation

Gases

Now we come to the problem of calculating the thermodynamic properties of gases. This differs from what we have done for solids primarily in the method of counting the stationary states of the assembly. In solids the molecules could be considered as distinguishable in the sense that, being localised, their average positions served to identify them. In gases this is no longer true and we must take account of the fundamental indistinguishability of the molecules of our assembly.

In considering this problem we make use of the fact that, under most of the conditions we deal with, the energy levels available to the gas molecules are very close together ($\Delta \varepsilon \ll kT$) and it is therefore convenient to lump them together in groups:

w_i levels with energy all very nearly ε_i
w_{i+1} levels with energy all very nearly ε_{i+1}
w_{i+2} levels with energy all very nearly ε_{i+2}

and so on.

The number of energy levels available to a gas molecule under most circumstances is enormous. For example in a mole of gaseous oxygen at atmospheric pressure and room temperature, the number of energy levels accessible to each molecule exceeds 10^{30}. Thus, even if we divide up the energy into $\sim 10^{10}$ different groups of levels, each group would still contain typically $\sim 10^{20}$ levels. Because of these very large numbers the grouping together can be done without serious error (except in certain special cases, of which more later). Moreover in general the w_is are very large numbers indeed which is useful for purposes of mathematical approximation.

115

We now consider an assembly of N molecules, of fixed volume V and fixed energy E. Take a typical distribution in which:

n_1 molecules are in a state of energy ε_1 consisting of w_1 levels
n_2 molecules are in a state of energy ε_2 consisting of w_2 levels
. . .
n_i molecules are in a state of energy ε_i consisting of w_i levels
. . .

(Notice that the occupation number n_i is not here the number of particles in the ith quantum level as before but the number of particles in the w_i group of levels which have been lumped together.)

We now concentrate on a typical set of levels, namely the w_i at energy ε_i which contain n_i molecules. Our first question is: in how many distinguishable ways can we assign indistinguishable molecules to the w_i distinct states? This is equivalent to asking: in how many distinguishable ways can we distribute n_i indistinguishable objects in w_i different boxes? The answer to this is known from the algebra of permutations and combinations. The number is (see Q1 of the exercises on this chapter).

$$p_i = \frac{(w_i + n_i - 1)!}{n_i!(w_i - 1)!} \tag{121}$$

Since we are assuming that both n_i and w_i are large numbers we can write this:

$$p_i = \frac{(w_i + n_i)!}{n_i! w_i!} \tag{122}$$

A possible microstate of the assembly is now obtained by combining any one of these arrangements with one each from all the other groups of stationary states. The total number of such microstates is thus

$$P = p_1 p_2 \ldots p_i \ldots = \prod_i \frac{(w_i + n_i)!}{n_i! w_i!} \tag{123}$$

We can quite well work with this expression and find the values of the n_is which make P a maximum. However, there is a further approximation (and simplification) which is justified in the majority of cases where we apply these results. This approximation arises because $w_i \gg n_i$. We know that

$$\frac{(w_i + n_i)!}{w_i!} = (w_i + n_i)(w_i + n_i - 1)(w_i + n_i - 2) \ldots (w_i + 1)$$

This contains n_i factors all of which are nearly equal to w_i if $n_i \ll w_i$. Thus we can write

$$\frac{(w_i + n_i)!}{w_i!} \simeq (w_i)^{n_i} \tag{124}$$

so that

$$p_i = \frac{(w_i)^{n_i}}{n_i!}$$

and

$$P = \prod_i \frac{(w_i)^{n_i}}{n_i!} \tag{125}$$

Taking logs and using Stirling's approximation we get

$$\ln P = \sum n_i \ln w_i - \sum (n_i \ln n_i - n_i) \tag{126}$$

The restrictive conditions are as usual:

$$\sum n_i = N$$

$$\sum n_i \varepsilon_i = E$$

Now we allow the n_is to vary and find the values which make $\ln P$ a maximum. For a maximum we require that

$$\sum \ln n_i \, \delta n_i - \sum \ln w_i \, \delta n_i = 0 \tag{127}$$

subject to

$$\sum \delta n_i = 0 \tag{128}$$

$$\sum \varepsilon_i \, \delta n_i = 0 \tag{129}$$

As before, we multiply equation (128) by α, and equation (129) by β and add them to (127). Thus:

$$\sum (\ln n_i - \ln w_i + \alpha + \beta \varepsilon_i) \, \delta n_i = 0 \tag{130}$$

If this is to be true for arbitrary values of the δn_is

$$\ln n_i^* - \ln w_i + \alpha + \beta \varepsilon_i = 0 \text{ for all } i$$

Therefore

$$n_i^* = w_i \, e^{-\alpha - \beta \varepsilon_i} \tag{131}$$

117

As before β can be identified as $1/kT$. Therefore

$$\frac{n_i^*}{N} = \frac{w_i e^{-\varepsilon_i/kT}}{\sum w_i e^{-\varepsilon_i/kT}} \tag{132}$$

Notice that the equilibrium occupation number n_i^* is proportional to w_i which means that the ratio n_i^*/w_i is independent of the value of w_i. Thus although the values of the w_i's were arbitrary, the average occupation number per quantum level (n_i^*/w_i) turns out to be perfectly well defined and free from ambiguity.

If you compare equation (132) with the corresponding equation (62) for a solid-like (or localised) assembly the two equations are essentially the same. Why then all this careful distinction between solids and gases, between localised and non-localised assemblies? The answer lies in equation (125). If you compare that with the corresponding result for a localised assembly (equation (53)) you will see that, apart from the factors $(w_i)^{n_i}$ (which arise from the grouping together of the levels), it differs by a factor of $N!$; $N!$ is in the numerator for the localised assembly and not for the gas. This is very important as far as the entropy is concerned. If the $N!$ were retained, the entropy of our gas would not be properly extensive, i.e. proportional to the mass (see equation (144) below). Boltzmann indeed recognised this failure of extensivity and dropped the $N!$ for this reason. Our method of counting which takes account of the fundamental indistinguishability of the molecules gets rid of the $N!$ quite naturally.

If we had not made the approximation involved in equation (125) but had used instead the more accurate equation (123) we should have reached the following result (as you may easily check):

$$n_i^* = \frac{w_i}{e^{\alpha + \beta \varepsilon_i} - 1} \tag{133}$$

instead of equation (131). This differs from equation (131) by having the additional term -1 in the denominator. Our original assumption that the n_i's were much smaller than the w_is was equivalent to the assumption that the exponential term in the denominator is very large compared to unity. Notice that if the equation has the form (133), i.e. if the 1 cannot be neglected, then it is *not* possible to eliminate α directly as we have done in equation (132). The distribution of energy defined by equation (133) is called a Bose–Einstein distribution after its discoverers. It is the equilibrium distribution appropriate to a gas of non-interacting identical molecules when no restrictions are imposed on the occupation numbers. Later we shall see that for some assemblies the occupation numbers must be limited to conform to the Pauli exclusion principle. For the present, however, we return to the simpler form of distribution (a classical distribution) as expressed in equations (131) and (132). We shall consider later when we are justified in using this approximation.

The entropy of an ideal gas

We assume that our fundamental relationship between entropy and the number of microstates of the assembly is valid also for gases so that the entropy of our gas is given by

$$S = k \ln W = k \ln P_{\max}$$

From equation (125) we can deduce $P_{\max}$ if we substitute in it the equilibrium distribution numbers, i.e. the n_i^*s. We then get (dropping the asterisks and remembering that we are now referring to the equilibrium distribution numbers)

$$\frac{S}{k} = - \sum (n_i \ln n_i - n_i) + \sum n_i \ln w_i$$

$$= - \sum n_i \ln n_i / w_i + N \tag{134}$$

Now use the result that

$$\frac{n_i}{w_i} = \frac{N e^{-\varepsilon_i / kT}}{\sum w_i e^{-\varepsilon_i / kT}} \tag{135}$$

and write

$$\sum w_i e^{-\varepsilon_i / kT} = Z \tag{136}$$

This is the partition function again (cf. equation (102)).
Substitute (135) in (134). Therefore

$$\frac{S}{Nk} = - \sum \frac{w_i e^{-\varepsilon_i / kT}}{Z} \left\{ \ln (N/Z) - \frac{\varepsilon_i}{kT} \right\} + 1$$

$$= \sum \frac{\left(\dfrac{w_i \varepsilon_i}{kT} \right) e^{-\varepsilon_i / kT}}{Z} - \ln N + \ln Z + 1 \tag{137}$$

Notice that in the sum the numerator is just $T(\partial Z / \partial T)$ as you can easily verify by differentiating Z with respect to T. Therefore

$$\frac{S}{Nk} = \frac{T}{Z} \frac{\partial Z}{\partial T} - \ln N + \ln Z + 1 \tag{138}$$

Thus to evaluate S all we need to know is the partition function, Z.

The evaluation of the partition function, Z, for an ideal gas

The partition function, Z, is

$$Z = \sum_i w_i e^{-\varepsilon_i/kT}$$

w_i, remember, is the number of stationary states with energy ε_i. We can conveniently transform the sum into an integral in the following way. Let $g(\varepsilon)\,d\varepsilon$ be the number of quantal stationary states with energies lying between ε and $\varepsilon + d\varepsilon$. Here $g(\varepsilon)$ is some function of ε which we must find once we know the allowed energy levels of the molecules of the gas. (We are now treating ε as a continuous variable, which is legitimate because the energy levels are so closely spaced.) Making this transformation we get

$$Z = \int_0^\infty g(\varepsilon)\, e^{-\varepsilon/kT}\, d\varepsilon \tag{139}$$

For our present purposes we are going to consider the molecules of the gas as point particles with no magnetic moment, no internal vibrations and no rotations; no energy, in fact, except that due to their moving about inside a box. For convenience we choose a cubic box of side L. Then according to quantum mechanics the energy levels allowed to a point particle of mass, m, confined in such a box are given by

$$\varepsilon_{p,q,r} = \frac{h^2}{2mL^2}(p^2 + q^2 + r^2) \tag{140}$$

where p, q and r are any positive or negative integers. To each set of integers there is one and only one stationary state.

With this information we now want to find $g(\varepsilon)\,d\varepsilon$, the number of stationary states with energy between ε and $\varepsilon + d\varepsilon$. To do this, let us rewrite equation (140) as

$$p^2 + q^2 + r^2 = \frac{2mL^2}{h^2}\,\varepsilon \tag{141}$$

We now treat p, q and r as Cartesian co-ordinates and, because they are large numbers in all our present applications, we treat them as continuous variables. For a given value of ε, the expression (141) is now the equation of a sphere of radius $(L/h)\sqrt{2m\varepsilon}$. Moreover, all states whose energy is *less* than ε must be represented by points lying inside the sphere. Within this sphere there is just one stationary state per unit volume because when either p or q or r increases by unity this generates a new set of integers corresponding to a new quantum state as defined by equation (140). Thus the number of states with energy less than ε is just equal to the volume of the sphere.

We can now find $g(\varepsilon)\,d\varepsilon$. The number of stationary states with energies between ε and $\varepsilon + d\varepsilon$ is just equal to the volume contained between the sphere corresponding to the energy ε and that corresponding to $\varepsilon + d\varepsilon$. This is the volume of the spherical shell lying between the radii r and $r + dr$, where $r = (L/h)\sqrt{2m\varepsilon}$.

The volume of this shell is $4\pi r^2\,dr$

$$= 4\pi \frac{L^2}{h^2} 2m\varepsilon \, d\left(\frac{L}{h}\sqrt{2m\varepsilon}\right)$$

Thus the number of stationary states with energies between ε and $\varepsilon + d\varepsilon$ is, in this problem,

$$g(\varepsilon)\,d\varepsilon = 4\pi \frac{L^3}{h^3}(2m)^{3/2}\frac{\varepsilon^{1/2}}{2}\,d\varepsilon \tag{142}$$

Thus

$$Z = \int_0^\infty g(\varepsilon)\,e^{-\varepsilon/kT}\,d\varepsilon$$

$$= \frac{2\pi L^3}{h^3}(2m)^{3/2}\int_0^\infty \varepsilon^{1/2}e^{-\varepsilon/kT}\,d\varepsilon$$

If we now make the variable $\varepsilon/kT = x$ instead of ε we get

$$Z = \frac{2\pi L^3}{h^3}(2mkT)^{3/2}\int_0^\infty x^{1/2}e^{-x}\,dx$$

The value of this definite integral is $\sqrt{\pi}/2$; therefore

$$Z = \frac{V(2\pi mkT)^{3/2}}{h^3} \tag{143}$$

where we have written V, the volume of the gas, instead of L^3 (it can be shown that the *shape* of the container does not alter the result).

The Sackur–Tetrode equation for the entropy of an ideal gas

We have now evaluated the partition function Z of our gas so we are in a position to calculate its entropy. We go back to equation (138) and substitute our value of Z. From this we get

$$\frac{S}{Nk} = -\ln\frac{N}{V} + \frac{3}{2}\ln\left(\frac{2\pi mkT}{h^2}\right) + \frac{5}{2} \tag{144}$$

This is a very important result for the entropy of an ideal gas of point particles; it is often called, after the two people who first derived it, the Sackur–Tetrode equation.

From it we can find at once that the heat capacity at constant volume:

$$C_V = T\left(\frac{\partial S}{\partial T}\right)_V = 3Nk/2 \tag{145}$$

For one mole of gas this means that $C_V = 3R/2$, which is constant and independent of temperature. (Incidentally, this would cease to be true at low enough temperature because the approximation expressed in equation (124) would cease to hold: see discussion on p. 131 and Figure 14.)

Now let us look at the expression for S in more detail; there are a number of points to notice:

(i)   The general form of the expression is consistent with the thermo-dynamic deduction of the entropy of a perfect gas having a constant heat capacity (see equation (27) Part One). Thus S (per mole) depends on $C_V \ln T$ and on $R \ln V$.

(ii) S is extensive, i.e. if the temperature T and the density of the gas, N/V, are kept constant S is proportional to the mass of gas, i.e. to N. As I emphasised earlier, if we had had the additional term $N!$ in the expression for P_{max}, this would have upset the extensivity of the result for S. This emphasises the importance of correctly counting the microstates in solids on the one hand and gases on the other.

(iii) Apart from the terms $\frac{3}{2}Nk \ln T$ and $Nk \ln V$ the other quantities in the expression for S are all constants for a fixed mass of gas. These form the so-called entropy constant. In this case it has the value

$$Nk\left\{\frac{3}{2}\ln\frac{2\pi mk}{h^2} - \ln N + \frac{5}{2}\right\}$$

We shall discuss the significance of this term when we discuss the third law of thermodynamics. At present it is sufficient to note that the value of the entropy at *high temperatures* obtained from the Sackur–Tetrode equation is correct; at low temperatures (the precise criterion will be given below) the equation is no longer valid.

Gas mixtures

The methods we have already used can readily be extended to deal with the entropy of a mixture of gases; unfortunately space does not allow me to discuss these questions here. I will, however, just mention one result. If we have several gases A, B, C, occupying the same volume V, the entropy of the mixture is given by:

$$S_{total} = S_A + S_B + S_C + \dots$$

where S_A, S_B, $\dots$, are the entropies that the gases A, B, $\dots$, would have if they occupied the same volume V separately.[10]

The distribution of velocities in the gas

Before going any further it is convenient at this point to use the information we already have to work out how the velocities of the molecules in the gas are distributed, i.e. how many molecules have velocities between, say, v and $v + dv$. (Note that this symbol is a 'vee', not a Greek 'nu'.)

We know from equation (132) that the number (n_i) of molecules with energy ε_i in a gas at temperature T is given by

$$\frac{n_i}{w_i} = \frac{N e^{-\varepsilon_i/kT}}{Z}$$

Let us rewrite this as a continuous function of the energy ε in terms of $g(\varepsilon)$ (equation (142)).

To make the change we do three things: we replace n_i by dn, the number of molecules having energy between ε and $\varepsilon + d\varepsilon$; at the same time we replace w_i by $g(\varepsilon)\,d\varepsilon$, the number of energy states between ε and $\varepsilon + d\varepsilon$; and we replace ε_i by the continuous variable, ε. We then get

$$dn = \frac{N g(\varepsilon)\, e^{-\varepsilon/kT}}{Z}\, d\varepsilon$$

Now substitute for $g(\varepsilon)$ from equation (142) and for Z from equation (143) and tidy up the expression. We then find

$$dn = \frac{2N}{\pi^{1/2}} \frac{\varepsilon^{1/2}\, e^{-\varepsilon/kT}}{(kT)^{3/2}}\, d\varepsilon$$

Now, since the energy of the molecules is entirely kinetic, we can write $\varepsilon = \frac{1}{2}mv^2$. We then get

$$dn = N\left(\frac{2}{\pi}\right)^{1/2}\left(\frac{m}{kT}\right)^{3/2} v^2 e^{-mv^2/2kT}\, dv \tag{146}$$

This is now an expression for the number of molecules with velocities between v and $v + dv$. It is called the Maxwell–Boltzmann distribution law; the form of the distribution has already been illustrated in Figure 7.

From this result it is then possible to deduce various quantities which characterise the velocity distribution, e.g. the mean speed, the root mean square velocity, the most probable velocity and so on. One very useful quantity is the mean kinetic energy of the molecules. This is given by

$$\tfrac{1}{2}m\overline{v^2} = \frac{\displaystyle\int_0^\infty \frac{mv^2}{2}\, dn}{\displaystyle\int_0^\infty dn} \tag{147}$$

123

This is readily shown to give

$$\tfrac{1}{2}m\overline{v^2} = \tfrac{3}{2}kT \tag{148}$$

We see therefore that the mean kinetic energy of a molecule is proportional to the absolute temperature, T, a result well known from the kinetic theory of gases. If the components of velocity are treated separately we find

$$\tfrac{1}{2}m\overline{v_x^2} = \tfrac{1}{2}m\overline{v_y^2} = \tfrac{1}{2}m\overline{v_z^2} = \tfrac{1}{2}kT \tag{149}$$

This is a particular case of a more general theorem in 'classical' statistical mechanics called the *principle of equipartition of energy*. According to this, each independent 'squared' term (here $\tfrac{1}{2}mv_x^2$, for example) in the expression for the mechanical energy of the assembly contributes $\tfrac{1}{2}kT$ to the thermal energy when the assembly is in equilibrium at temperature T. This holds only in the 'classical' temperature region in which kT is large compared to the spacing between the relevant quantal energy levels.

This theorem of the equipartition of energy which was first demonstrated by Boltzmann was at one time used as an argument against his theories. The theorem implies that if a gas has atoms with 'internal degrees of freedom' (i.e. if the atoms themselves have a structure) these internal degrees of freedom will take up thermal energy and contribute to the specific heat. Already in Boltzmann's time it was known from spectra that atoms must have some sort of internal structure, so it followed from the equipartition law that this structure should show itself in an additional contribution to the heat capacity of gases. However, experimentally no such contribution was found.

The contradiction was resolved by the quantum theory. The 'structural' energy states of atoms are quantised and a typical energy difference between the ground state and the first excited state, $\Delta\varepsilon$, is usually so large that at any temperature, T, of normal measurement, kT is very much less than $\Delta\varepsilon$; so these upper states are not excited and do not contribute to the heat capacity.

However, in Boltzmann's day the quantum theory had not been formulated so that the kind of argument outlined above provided powerful evidence against his views. Towards the end of his life Boltzmann became pessimistic and severely depressed, torn, no doubt, between a conviction that he was right and an inability to prove it. In 1906 he committed suicide. It is ironical that by this time his great contributions to physical theory had been largely vindicated and had provided the foundation on which Planck built his theory of quanta. It was this theory which ultimately made it possible to resolve many of the paradoxes that had plagued Boltzmann's own work.

The Helmholtz free energy

The Helmholtz free energy, F, is defined as

$$F = E - TS$$

From equation (105), which holds for both solid-like and gas-like assemblies, and equation (138) we get

$$F = -NkT \ln Z + NkT \ln N! \tag{150}$$

We will not evaluate this but will use this result to find the equation of state of the gas.

The equation of state

From thermodynamics we know that

$$p = -\left(\frac{\partial F}{\partial V}\right)_T \tag{151}$$

so here, from (150),

$$p = +NkT \frac{\partial \ln Z}{\partial V} \tag{152}$$

and from equation (143)

$$p = + \frac{NkT}{V}$$

Therefore

$$pV = NkT \tag{153}$$

is the equation of state of our assembly.

If we have one mole of a perfect gas its equation of state is

$$pV = RT$$

where R is the gas constant per mole. If we compare the last two equations we see that if our assembly contains one mole of gas (i.e. if $N = N_0$, Avogadro's number) then

$$N_0 k = R \tag{154}$$

as we assumed in our earlier work.

We see therefore that our model of a gas reproduces all the expected thermodynamic properties of an ideal gas; on the other hand the statistical treatment gives a great deal more information (notably about the entropy) than does a purely thermodynamic treatment.

Exercises

Q1 Show that the number of different ways of assigning n_i indistinguishable objects in w_i boxes is given by:

$$p_i = (n_i + w_i - 1)!/n_i!(w_i - 1)! \tag{Equation (121)}$$

125

(Hint: Represent each object by a circle ○ and lay n_i of them in a single continuous line. Now represent the boxes by inserting vertical bars | at appropriate places to split off the desired number of objects thus:

○○○○|○○|○○○○||○ . . . . |○○○|○|○○○○○

which would correspond to a distribution with four objects in the first box, two in the second, four in the third, none in the fifth and so on. Notice that there are only $\omega_i - 1$ bars, because although each bar creates a box to its right, the first bar also creates a box to its left. Now calculate the number of different ways you can do this.)

Q2 (a) What is the total change in entropy when 1 mole of hydrogen at a pressure p_1 of 1 atmosphere mixes adiabatically with 3 moles of hydrogen at $p_2 = 3$ atmospheres and the same temperature? The total volume is unchanged in the process. (b) What would be the total change of entropy in analogous circumstances if the gas at 3 atmospheres was helium instead of hydrogen?

Q3 The quantum energy levels of an object rotating about a given axis of rotation, whose moment of inertia about that axis is I, are given by:

$$\varepsilon_j = h^2 j(j+1)/8\pi^2 I$$

where $j = 0, 1, 2, 3 \ldots$ The symmetry of the object can limit the allowed transitions but the size of possible energy changes is given approximately correctly. How does this explain why a monatomic gas shows no rotational contribution to its specific heat capacity and why a gas of diatomic molecules shows a contribution from only two principal axes of rotation? (The interatomic separation in a diatomic molecule is of the order of 10^{-10} m while the diameter of the nucleus is of order 10^{-17} m. The mass of the proton is of order 10^{-27} kg and that of the electron roughly 2000 times less, $h \approx 6.6 \times 10^{-34}$ J s; $k \approx 1.4 \times 10^{-23}$ J K^{-1}.)

Application to gases: (2) Bose–Einstein and Fermi–Dirac gases

A Fermi–Dirac assembly

Before we discuss the conditions under which the results of the preceding chapter are valid it is useful first to derive some properties of a different kind of gas. This is a gas of particles that obey the Pauli exclusion principle; such an assembly is said to obey Fermi–Dirac statistics (again named after the discoverers) as opposed to the Bose–Einstein statistics we have already considered. Examples of gases that would obey Fermi–Dirac statistics are a gas of He^3 atoms or an electron gas.

If the Pauli exclusion principle applies to the particles we are dealing with, then only *one* particle can occupy a particular quantum state. Consequently our total number of stationary states for the assembly has to be recalculated taking this into account. As before, we group the quantum energy levels into groups and consider N particles distributed as follows:

n_1 particles in the w_1 states of energy close to ε_1

n_2 particles in the w_2 states of energy close to ε_2

. . .

n_i particles in the w_i states of energy close to ε_i

. . .

Now we concentrate on the n_i particles which are distributed among the w_i states with energy ε_i. Notice that now w_i must be greater than n_i because we cannot have more than one particle per state. We now wish to calculate in how many ways we can assign n_i indistinguishable particles into w_i states with no more than one particle per state. In other words, in how many ways can we put n_i indistinguishable objects into w_i distinct boxes with no more than one in each? We could imagine the w_i boxes in a row and we could describe a particular distribution by a sequence of w_i 0s or 1s. A zero means the box is empty: a one means the box is full. There are thus n_i 1s and $(w_i - n_i)$ 0s. In how many distinguishable ways can we write down a row containing n_i 1s and $(w_i - n_i)$ 0s? The answer is

$$^{w_i}C_{n_i} = \frac{w_i!}{n_i!(w_i - n_i)!} \tag{155}$$

Thus the number of distinguishable stationary states associated with the n_i particles with energy ε_i is

$$p_i = \frac{w_i!}{n_i!(w_i - n_i)!}$$

and as each of these can be combined with any of the stationary states from the other levels, the total number of stationary states for the assembly is

$$P = p_1 p_2 \cdots p_i \cdots = \prod \frac{w_i!}{n_i!(w_i - n_i)!} \tag{156}$$

Now if we maximise P subject to the restrictive conditions

$$\sum n_i = N$$

$$\sum n_i \varepsilon_i = E$$

we find quite easily that

$$n_i = \frac{w_i}{e^{\alpha + \beta \varepsilon_i} + 1} \tag{157}$$

This is the Fermi–Dirac distribution function. It differs from the Bose–Einstein distribution (equation (133)) because the denominator contains now +1 not −1. If in equation (156) we had assumed that $w_i \gg n_i$ we could, by an approximation similar to the one we made in equation (124), have written

$$P \simeq \prod_i \frac{(w_i)^{n_i}}{n_i!} \tag{158}$$

which is just the same as equation (125) and leads to the so-called 'classical' distribution function expressed by equation (131) or (132). Let us compare the three distribution functions:

Bose–Einstein	Fermi–Dirac	Classical	
$n_i = \dfrac{w_i}{e^{\alpha + \beta \varepsilon_i} - 1}$;	$n_i = \dfrac{w_i}{e^{\alpha + \beta \varepsilon_i} + 1}$;	$n_i = \dfrac{w_i}{e^{\alpha + \beta \varepsilon_i}}$	(159)

The third is an approximation to both of the first two when $e^{\alpha + \beta \varepsilon_i}$ becomes very large compared to unity.

When do we use these different distribution functions? We answer this question in the next section.

The different distribution functions

The Fermi–Dirac distribution applies to assemblies of particles which are subject to the Pauli exclusion principle. Examples of such particles are: electrons, protons, neutrons and composite particles containing an odd number of protons, neutrons and electrons, e.g. He^3 atoms and so on.

The Bose–Einstein distribution applies to assemblies of the remaining kinds of particles such as light quanta (photons) and He^4 atoms.

Notice that the *distribution functions* apply to any sets of energy levels whatever, provided only that the particles can be treated as effectively independent. When we discuss the properties of an ideal Bose–Einstein or Fermi–Dirac *gas* we have to go a step further and specify the energy levels. The allowed translational energy levels of an ideal gas molecule are the quantum levels of a particle in a box as given in equation (140) above.

It then turns out, as I have already mentioned, that over a wide range of densities and temperatures the differences between the ideal Bose–Einstein, Fermi–Dirac and classical gases are negligible. Why is this, and what are the conditions of density and temperature that make it true? These questions can be answered if we recall the original reason for introducing the different statistics. The Bose–Einstein statistics arose from a consideration of assemblies in which any number of particles could exist in the same quantum state. Fermi–Dirac statistics on the other hand were developed to deal with assemblies in which at most only one particle was allowed in each quantum state. The enumeration of the distinguishable states of the assembly was then different in the two cases.

However, now suppose that the temperature and density in the assembly are such that there are accessible to the particles very many more quantum states than there are particles. Then the probability of finding two particles in the same quantum state will become exceedingly small so that the restriction, imposed by the Fermi–Dirac statistics, of one particle per state will make practically no difference. We can therefore expect that under these circumstances the difference between Fermi–Dirac and Bose–Einstein statistics will disappear.

Now let us see if we can give a numerical criterion for this state of affairs. To do this, let us first find out what spread of energy would be produced if we put just one particle into each of the N lowest energy levels available to the particles of the gas. (These energy levels are the ones specified by equation (140).) If we have N particles in an assembly of volume V it is not

129

difficult to show that in these circumstances all the levels up to E_0 will be occupied where E_0 is given by[11]

$$E_0 = \frac{h^2}{2m}\left(\frac{3N}{4\pi V}\right)^{2/3}$$ (160)

The symbols here have the same sense as earlier in this chapter. This expression implies that E_0 will be large if the mass of the particles m is small and if the number of particles per unit volume N/V is large.

At a high temperature T the average energy per particle in the assembly is of order kT, so we can expect *a priori* that any energy level between 0 and about kT will be available to any particular particle. Therefore, if $kT \gg E_0$ the number of energy states available to each particle will be so large that the chance of finding two particles in the same level will be very small indeed. Thus the condition that the difference between a Bose–Einstein and a Fermi–Dirac gas should be negligible is that kT should be very large compared to E_0. If we put in the value of E_0 and rearrange the terms we can express this condition as

$$\frac{3N}{4\pi V} \frac{h^3}{(2mkT)^{3/2}} \ll 1$$

This requirement is more usually written in the following form which is identical to the other except for a small numerical factor of no consequence:

$$\frac{N}{V} \frac{h^3}{(2\pi mkT)^{3/2}} \ll 1$$ (161)

This condition will thus tend to be satisfied if the number of particles per unit volume N/V (i.e. the density of the gas) is small, if the temperature is high and if the mass, m, of the particles is large.

If the condition (161) is *not* satisfied then we must use the appropriate quantum statistics (either Bose–Einstein or Fermi–Dirac) and work out the properties of the gas without making the 'classical' approximation.

Let us evaluate the quantity

$$\frac{N}{V} \frac{h^3}{(2\pi mkT)^{3/2}}$$

(which is sometimes called the 'degeneracy parameter' of the gas) for a special case. Helium is the substance with the lowest boiling point. At 1 K, He^4 (this is the heavier stable isotope of helium) has a vapour pressure of about 10^{-2} mm mercury, and so the molar volume of the vapour is $\sim 10^7$ cm^3. Consequently the degeneracy parameter for helium vapour in this state turns out to be $\sim 10^{-4}$. Degeneracy effects are thus small and easily masked by the bigger effects that arise because the vapour is not a perfect gas, i.e. because the helium atoms do in reality interact. For gases other than helium degeneracy effects are still smaller and for real gases the classical statistics are almost always adequate. There are, however, systems which behave like gases (e.g. the so-called electron gas in a metal) where the quantum statistics are needed. We shall therefore look briefly at the behaviour of an ideal Bose–Einstein and Fermi–Dirac gas at low temperatures although without working out any details.

Some properties of a Bose–Einstein gas at low temperatures

We have already seen how the Bose–Einstein and the Fermi–Dirac distributions go over to a common 'classical' distribution at high temperatures. What happens at low temperatures?

Since in a Bose–Einstein gas there is no restriction on the number of particles which can occupy any energy state it is clear that at 0 K all the particles must be in the lowest level in order to secure the minimum total energy of the assembly. This state of affairs can be achieved in only one way since the particles are to be thought of as indistinguishable, so that permuting them within a given level produces no new microstates. Thus $W = 1$ and $S = k \ln W$ is zero. Therefore the entropy of the gas at 0 K is zero.

For temperatures above 0 K it is possible to work out the thermodynamic properties in detail from the distribution function and the energy levels;[12] it turns out that the Bose–Einstein gas behaves in a peculiar way as the temperature tends to zero at constant density. As the particles begin to 'drop' into the ground state an effect similar to condensation occurs. This shows itself in the shape of the isotherms which become horizontal at a certain density for each temperature. This is referred to as Bose–Einstein condensation and for each density there is a critical temperature T_c at which the effect begins and at which the heat capacity undergoes a sudden change of slope (see Figure 14). The temperature T_c is related to the volume, V, and number of particles in the gas, N, by the relationship:

$$T_c = \frac{h^2}{2\pi mk}\left(\frac{N}{2.612V}\right)^{2/3} \tag{162}$$

T_c is often called the degeneracy *temperature* of the gas. Clearly it is closely related to the degeneracy *parameter* defined in equation (161); indeed T_c is the temperature at which the degeneracy parameter of the gas equals 2.612 (this curious number arises from a definite integral that appears in the detailed working out of the theory).

To gain a clearer insight into the physics involved here we can compare the de Broglie wavelength of the particles at T_c with the average distance between particles $(V/N)^{1/3}$. The de Broglie wavelength is defined in terms of the momentum mv of the particle as: $\lambda = h/mv$. The kinetic energy corresponding to this momentum is $mv^2/2$, which expressed in terms of λ is

$h^2/2m\lambda^2$. At temperature T_c, the average kinetic energy of a particle is roughly $3kT_c/2$ and so from equation (162), we can write for the value of λ at T_c:

$$\frac{3h^2}{4\pi m}\left(\frac{N}{2.612V}\right)^{2/3} \simeq \frac{h^2}{2m\lambda^2}$$

or

$$\frac{V}{N} \simeq \frac{\lambda^3}{8}$$

This indicates that at the critical temperature for Bose–Einstein condensation there are quite a number of particles lying within a volume of λ^3 defined by their de Broglie wavelength. Their wavefunctions thus strongly overlap. The Bose–Einstein condensate is then no longer a set of independent particles: instead the wavefunctions of the individual particles become coherent with those of their neighbours and we now have collective behaviour of the whole condensate.

The difficulty in observing Bose–Einstein condensation is that it tends to be masked by the mutual interaction between particles. Let us suppose that this interaction has a range r_0 so that at distances greater than this the interaction is negligible. If then $\lambda \ll r_0$, the particles are dominated by the interaction and quantum effects can be ignored. If, however, $\lambda \gg r_0$, then the quantum statistics will dominate the behaviour. Thus the choice of element and hence r_0 determines the degree of cooling (and hence the value of λ) that has to be achieved.

The phenomenon of Bose–Einstein condensation has been seen in a very weakly interacting collection of rubidium atoms cooled to a kinetic temperature of about 10^{-7} K by laser cooling (see below) and held in a suitable magnetic trap. To demonstrate the effect we need to know how the velocities of the atoms in the condensate are distributed. This can be inferred from the shadow cast when the condensate is suitably illuminated; the precise distribution depends, of course, on the details of the trap but the existence of Bose–Einstein condensation could, in fact, be inferred from the experimental data in this way.[13]

The principle behind laser cooling is that atoms that absorb a photon are given a kick in the direction of the photon. Thus if a beam of atoms is directed towards a laser beam whose frequency ω_L differs slightly from the frequency of an atomic transition ω_A, those atoms whose velocity is just right to bring the Doppler shifted frequency into resonance will tend to absorb photons from the laser. The photons emitted by the excited atoms go in random directions so that there is a general slowing down of the atoms in the laser direction. By suitable arrangements of the frequency and geometry of the lasers this method can be used to slow atoms sufficiently to hold them in a magnetic trap. Further cooling can then be achieved because particles of higher than average energy can be ejected from the trap, so further cooling the remainder.

Hydrogen in gaseous form tends to exist only as diatomic molecules, but in 1979 it was found that atomic hydrogen could be kept for several minutes at low temperatures in a container whose walls were coated with a film of liquid helium. Although not yet fully realised, atomic hydrogen has great potential as a quantum system because of its high degeneracy temperature and small range of interaction r_0.

Bose–Einstein condensation is also thought to be involved in the strange properties not of a gas, but of a liquid, liquid He^4, as we shall discuss in Chapter 14.

Some properties of an ideal Fermi–Dirac gas

In a Fermi–Dirac assembly only one particle may occupy each non-degenerate energy level. Some of the simpler consequences of this for an ideal gas can be deduced without doing a detailed calculation of its properties.

Consider first the absolute zero. The minimum value for the energy of the gas is obtained when all the lowest allowed translational energy levels are filled with particles (one particle to each level). Thus if there are N particles, the lowest N energy states are occupied. We have already seen that under these conditions all the energy states up to a value E_0 are occupied where E_0 is given by the expression:

$$E_0 = \frac{h^2}{2m}\left(\frac{3N}{4\pi V} \right)^{2/3} \tag{112}$$

E_0 is the energy of the highest occupied state so that the average energy per particle at 0 K must be less than this. If the energy levels were distributed uniformly in energy the mean energy would be $E_0/2$ but because the number of levels in a given energy range varies as $\sqrt{E}$ (see equation (142)), the mean energy at 0 K turns out to be $3E_0/5$. This can produce a very appreciable zero-point energy for the assembly if the density of the gas is high (i.e. if the energy levels are significantly separated from each other).

Since at 0 K there is precisely one particle in each of the lowest N quantum states it is clear that there is one and only one way of achieving this arrangement with indistinguishable particles. Therefore $W = 1$ and $S = 0$. Thus at the absolute zero the assembly has zero entropy.

At very high temperatures (for a given volume $kT \gg E_0$) we know that the Fermi–Dirac and Bose–Einstein distributions go over to the classical one and all three have the same expression for the entropy, already given in equation (144).

At low temperatures but above the absolute zero we can also say something about how the Fermi–Dirac gas behaves; suppose that $kT \ll E_0$. It is clear that the restriction whereby we can have only one particle per quantum level severely limits the effect of a small amount of thermal energy. Those particles which are in low-lying energy states (i.e. those whose energy differs from E_0 by some quantity much greater than kT) will not be affected by the thermal energy for the following reason. At temperature T, each particle may receive on the average an energy of about kT in addition to its zero-point energy. However, if there are no empty levels for the particle to go into at the new value of the energy, it cannot receive this additional thermal energy. If it did it would have to go into a level already occupied which, in such an assembly, is forbidden by the exclusion principle.

Thus only those particles which lie within an energy range of about kT of E_0 can accept the thermal energy. The *fraction* of particles affected is thus $\sim kT/E_0$ and so the actual *number* of particles affected is $\sim NkT/E_0$. Since each of the excited particles receives an additional energy of kT the thermal energy of our assembly at temperature T is

$$U_T \sim N\left(\frac{kT}{E_0}\right)kT \tag{163}$$

So the heat capacity, C_V, is given by

$$C_V = \left(\frac{\partial U_T}{\partial T}\right)_V \approx 2Nk\left(\frac{kT}{E_0}\right)$$

in the temperature region where $kT \ll E_0$. If the detailed calculations are made these show that the true expression for C_V is

$$C_V = \frac{\pi^2}{2}Nk\left(\frac{kT}{E_0}\right) \tag{164}$$

The important point is that C_V is linear in T at these temperatures and of order $R(kT/E_0)$ for one mole.

If we deduce the entropy from the heat capacity we get

$$S = C_V = \frac{\pi^2}{2}Nk\left(\frac{kT}{E_0}\right) \tag{165}$$

The internal energy and heat capacity of an ideal Fermi–Dirac gas are shown as a function of reduced temperature (T/T_c) in Figures 12 and 14. (T_c is defined in equation (162).) For comparison the same quantities are also plotted for a Bose–Einstein and classical gas at the same density. In Figure 13 a comparison is made between the translational entropy of a Fermi–Dirac gas and that of a Bose–Einstein gas at the same density. You see that both curves start from zero at 0 K but in different ways. You also see that at high temperatures both curves merge into each other and into that of a 'classical' gas which is also shown in the figure. The entropy of the classical gas is calculated from the Sackur–Tetrode equation; the entropy so calculated diverges at 0 K and goes off to minus infinity. However, the important point here is that although the Sackur–Tetrode equation fails at low temperatures

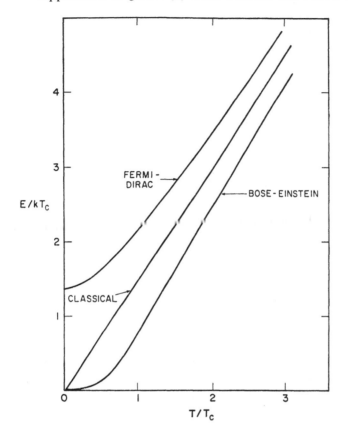

Figure 12 Energy of Bose–Einstein, Fermi–Dirac and classical gas (all at the same density) as a function of the reduced temperature (T/T_c).

nevertheless at high temperatures (the criterion for 'high' and 'low' has already been discussed) it gives correctly the entropy of a gas with reference to its true value at $0\,K$.

All real gases, insofar as they approximate to ideal gases, must be either ideal Bose–Einstein or ideal Fermi–Dirac gases or mixtures of the two. The Sackur–Tetrode equation is valid where, and only where, it gives the same answers as the correct Bose–Einstein or Fermi–Dirac treatment.

Exercises

Q1   A gas consists of three atoms with access to three distinct quantum states of the same energy. How many distinguishable microstates of this gas can be formed from these quantum levels for:

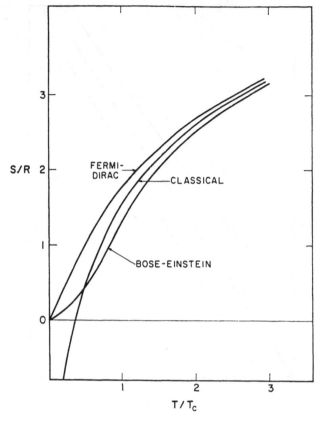

Figure 13 The entropy S/R as a function of the reduced temperature (T/T_c) for a Bose–Einstein, Fermi–Dirac and classical gas, all at the same density.

(a)   a classical gas, in which the atoms are distinguishable;
(b)   a Fermi–Dirac gas, in which the atoms are indistinguishable but only one atom can be in each state;
(c)   a Bose–Einstein gas in which the particles are indistinguishable but there is no restriction on occupation number.

Q2   What is the root mean square speed of a collection of atoms of atomic weight 85 whose kinetic temperature is 10^{-7} K?

Q3   The Fermi energy of the conduction electrons in metallic lithium is 4.7 eV measured from the bottom of the conduction band, Approximately what fraction of these electrons are excited thermally at 300 K? Estimate roughly the specific heat capacity of 1 mole of these electrons at 10 K.

Q4   What is the ratio of the Fermi energy of a fully degenerate gas of

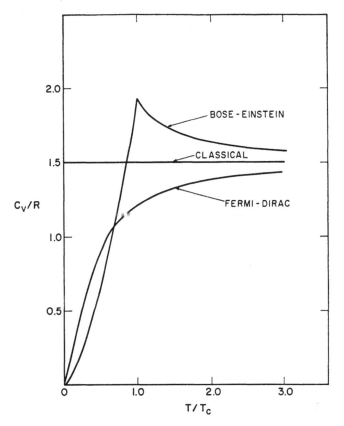

Figure 14 The heat capacity as a function of reduced temperature (T/T_c) of a Fermi–Dirac, Bose–Einstein and classical gas, all at the same density.

non-interacting electrons to that of a similar gas of non-interacting neutrons in the same volume? Why is this important in the formation of neutron stars? (The ratio of the mass of the neutron to that of the electron can be taken as 2000.)

13

Fluctuations and Maxwell's demon

What we have said so far about statistical mechanics applies strictly only to assemblies whose energy, E, and volume, V, are fixed and in which the number of particles, N, is fixed. Suppose, however, that we take an assembly with given values of E, V and N and place it in a constant temperature bath at the same temperature as that of the assembly. We now allow the assembly to exchange energy with the bath (i.e. we replace the condition that E is to be constant by the condition that the assembly is to be maintained at an equivalent constant temperature). Then it can be shown that the thermo-dynamic properties of the assembly are essentially unchanged provided that N is large, i.e. that the assembly is of large-scale dimensions in the thermodynamic sense. Likewise, if we replace the condition of constant volume by the condition of constant pressure (with the pressure so chosen as to maintain the density in the assembly unchanged), again we find that for large assemblies the thermodynamic properties are not altered. Indeed, it is not difficult to formulate a statistical treatment to deal with assemblies at constant temperature or constant pressure or constant chemical potential instead of at fixed E, V and N, and from this treatment to show that their properties under corresponding conditions are identical, if N is large enough.

If, however, we consider smaller assemblies or small parts of a large assembly we can expect to find that the properties of these small systems are not then the same under these different conditions. In fact, we may then expect that certain properties which would be effectively independent of time in the large assemblies become time dependent in the small ones. For example, suppose that we have a gas at a certain temperature and we mark off in space a certain small volume. If the volume is small enough the number of particles inside it will change appreciably from instant to instant and we shall have density fluctuations. Likewise the amount of energy enclosed by the volume will vary from instant to instant and we shall have energy fluctuations. If the volume we are considering becomes so small that it is comparable to the volume per molecule of the assembly then clearly sometimes it may be entirely empty and sometimes it may contain several molecules. Under these circumstances the fluctuations are very large indeed. Over a long period of time the average density and energy in the small volume will, of course, be the same as in the whole gas.

Other examples of fluctuations are the irregular spontaneous movements of tiny particles of material suspended in a liquid. The irregular movements arise because the particles in the liquid are not subject to a uniform pressure from all sides; because they are so small sometimes more and sometimes fewer molecules strike them from one side than from the other. This causes the particles to move about in the liquid quite randomly. These movements can be observed with a suitable microscope and are called 'Brownian movement' after the botanist Brown who first observed them. Noise in electrical circuits which arises from the random thermal motion of electrons in the conductors composing the circuits is another example of fluctations.

Statistical mechanics can make some predictions about fluctuations of this kind in small assemblies under equilibrium conditions. In general these predictions refer to the total mean square deviation of the quantity (for example, density or energy) from the mean value. If you try to measure fluctuations, however, you find that the magnitude of the departures from the mean depends on the time response of the apparatus you use to make the observation. For example, if you try to measure the fluctuations of electrical potential across the ends of an electrical conductor maintained at constant temperature the mean square fluctuations of voltage depend directly on the band width of the instrument used to measure them. Generally speaking, therefore, measurements of fluctuations take us into the realm of the *time dependence* of the properties of a system and lead us quite beyond the scope of conventional statistical mechanics.

A different aspect of fluctuations was considered by Maxwell in relation to the second law of thermodynamics. Imagine a gas in a vessel at a uniform temperature. 'Now let us suppose', says Maxwell,[14] 'that such a vessel is divided into two portions, *A* and *B*, by a division in which there is a small hole, and that a being, who can see the individual molecules, opens and closes this hole, so as to allow only the swifter molecules to pass from *A* to *B*, and only the slower ones to pass from *B* to *A*. He will thus, without expenditure of work, raise the temperature of *B* and lower that of *A*, in contradiction to the second law of thermodynamics.' This 'small but lively' being which was invented by Maxwell was called by Thomson (Lord Kelvin) a 'demon' and is nowadays referred to as Maxwell's demon.

From this argument Maxwell sought to demonstrate that the second law of thermodynamics has only statistical certainty. However, Brillouin,[15] following earlier work by Szilard, has shown that if the demon himself is subject to the laws of physics (in particular the quantum theory) the operations necessary to detect the fast molecules cause an increase in entropy sufficient to offset the decrease in entropy which the demon is trying to bring about in the gas. This kind of argument is probably quite general and would therefore lead to the following generalisation: it is impossible, even in principle, to exploit fluctuations in a systematic way so as to violate the second law of thermodynamics.

The important point here is that we cannot use fluctuations *systematically* to violate the second law. However, it is possible in principle that large fluctuations could occur purely by chance, even fluctuations that are large enough to bring about macroscopic changes. However, these are so improbable that for almost all purposes we can regard them as impossible. In principle, however, they could occur and that is why it is correct to say that the second law of thermodynamics has only statistical certainty.

The recurrence paradox

Boltzmann, as we have seen, showed how thermodynamic properties, in particular entropy, could be deduced from the mechanical properties of assemblies. The laws of mechanics, however, show that any closed conservative mechanical system must in time return to a state close to its (arbitrary) starting condition; moreover in principle it can do this indefinitely often. So it follows from our mechanical interpretation of entropy that the entropy too must undergo cyclic changes and so cannot, as we assume in thermodynamics, continually increase. This is the so-called *recurrence paradox*.

The paradox is resolved by recognising that in principle the entropy of an isolated system can indeed *decrease*, that it, too, is subject to fluctuations. However, when we consider the probabilities involved it is soon clear why we never encounter violations of the second law. Boltzmann, after emphasising that spontaneous changes from disordered to ordered states are conceivable, put the matter thus:[16]

> One should not however imagine that two gases in a $\frac{1}{10}$-litre container, initially unmixed, will mix, then separate again after a few days, then mix again, and so forth. On the contrary, one finds ... that not until after a time enormously long compared to $10^{10^{10}}$ years will there by any noticeable unmixing of the gases. One may recognise that this practically equivalent to *never*, if one recalls that in this length of time, according to the laws of probability, there will have been many years in which every inhabitant of a large country committed suicide, purely by accident, on the same day, or every building burned down at the same time – yet the insurance companies get along quite well by ignoring the possibility of such events.

Entropy and its physical meaning

Exercises

Q1 Give two examples of (a) macroscopic parameters that characterise large-scale transport properties of a physical system (the transport can be of heat, material or electricity); (b) microscopic quantities that characterise such properties. (Hint: think, for example, of quantities in the kinetic theory of gases.) (c) Contrast the latter with the microscopic quantities that characterise the statistical mechanics of a system.

Entropy at low temperatures

14

The third law of thermodynamics

Entropy at low temperatures

The third law of thermodynamics shows that the absolute zero is the natural origin from which to measure entropy. Consequently the concept of entropy is closely bound up with the approach to absolute zero. For this reason we shall first consider how the law can be stated and what it implies and then turn to the problem of producing and measuring low temperatures.

The third law in its earliest form was formulated by Nernst (1864–1941) in 1906 (it is sometimes called the Nernst heat theorem) and has subsequently been refined and clarified, notably by Simon (1893–1956). Nernst was very proud of his achievement in formulating what he referred to as 'my law'. He noted that there were three people associated with the discovery of the first law, two with the second and only one with the third. From this he deduced that there could be no further such laws.

For a long time the third law was a highly controversial topic. For example, here is a quotation from a paper of 1932:

> We reach therefore the rather ruthless conclusion that Nernst's Heat Theorem strictly applied may or may not be true, but is always irrelevant and useless – applied to 'ideal solid states' at the absolute zero which are physically useful concepts the theorem though often true is sometimes false, and failing in generality must be rejected altogether. It is no disparagement to Nernst's great idea that it proves ultimately to be of limited generality.'
> (Fowler and Stern on Statistical Mechanics and Entropy.)

Or as the Ruhemanns[1] described the situation: 'In the controversies which ensued the opponents of the third law gradually separated into two groups:

those who held that the law was true but inapplicable and those who maintained that though applicable it was false.'

The third law is no longer controversial; now that it is more fully understood the relative importance of different aspects of it has changed. Nevertheless it remains a valuable unifying principle and a useful guide in thinking about low temperature phenomena.

Let me state the law in its most unqualified form: *the entropy of all systems and of all states of a system is zero at absolute zero.*

Stated thus baldly, it would give rise to all the old arguments and criticisms. As we shall see, it has to be qualified to make it accurate, universal and useful, but this gives the essential meaning of the law.

In this statement the important thing is not that the value of the entropy at 0 K is *zero* but that it is the *same* for all systems or states of a system. You could give this entropy any value you pleased but since it would have to be the same for all systems and since its value would be entirely arbitrary, it is obviously simplest to put it equal to zero. We shall now look at the implications of this statement of the third law and see how it must be modified.

In discussing the statistical approach to entropy we noted that there were many different contributions to the *energy* of molecules: translational, vibrational, rotational, electronic, magnetic and so on. In a similar way we may think of separate contributions to the *entropy* of an assembly of molecules. Insofar as these contributions to the energy of an assembly are independent of each other, the corresponding contributions to the entropy of the assembly may be thought of separately. In this way we are led to talk about the vibrational entropy, the configurational entropy or the entropy of mixing of an assembly. If the assembly we are thinking of contains different kinds of particle – for example, the different kinds of atoms in an alloy or mixture of gases, or the electrons and ions in a metal – these too may sometimes be thought of as contributing separately to the total entropy. Thus we may have the nuclear spin entropy or the entropy of the conduction electrons in a metal and so on.

The third law can be applied to each of these contributions separately and this is very important. As we shall soon see, the positive assertions of the third law can only be applied to states between which reversible changes are possible (this is similar to the condition for being able to measure the entropy difference between these states). There are, however, many changes that cannot be made reversibly at low temperatures and about which the third law can in consequence make no positive assertion. However, because the entropy contributions can be separated, the law can still be applied to other properties of the same system where reversible changes *are* possible. Examples will make this clear below, but first let us consider how the entropy of some systems (in which reversible changes are possible) tends towards zero as the temperature approaches absolute zero. We shall consider only an outline of their behaviour in qualitative terms but this is enough to show how varied and sometimes how astonishing this behaviour is.

Superconductivity

Many metals and alloys abruptly lose all their electrical resistance at low temperatures and are said to become *superconducting*. Moreover the resistance can be restored by the application of a suitable magnetic field. In so-called type I superconductors, there is not only the vanishing of the resistance at the transition temperature T_c, but also, if the metal is in a magnetic field, the expulsion of all magnetic flux from inside the metal. This second effect, which is named after its discoverer and called the Meissner effect, shows that this superconducting state is a true thermodynamic state, which is independent of the path by which the metal arrives at it. In this it differs from a metal which has only the property of zero resistance, for which the magnetic state does depend on the path that leads to it.

Other metals or alloys (known as type II superconductors) show more complex behaviour and may show only a partial Meissner effect. More recently ceramic based materials have been found to show superconductivity at much higher temperatures than conventional superconductors. Some transition temperatures are shown for comparison in Table 8.

How does superconductivity come about? There may be more than one mechanism but a theory which explains what I have called conventional superconductivity was formulated in 1957 by Bardeen, Cooper and Schrieffer, (now called the BCS theory). The basic idea in this is that the conduction electrons interact with the lattice vibrations in such a way that pairs of electrons (Cooper pairs) are coupled together and in cooperation with other such pairs can reduce the free energy of the electrons below that of the comparable normal electron gas.

In simplified classical terms, the mechanism is as follows: an electron moving through the lattice of positive ions attracts to itself neighbouring ions through the mutual Coulomb attraction and sets them in motion towards the

Table 8 Some superconducting transition temperatures

Substance	T_c (K)
(conventional)	
Al	1.14
Hg	4.15
Pb	7.19
Nb	9.50
$NbSn_3$	18.05
(High T_c)	
$La_{2-x}Sr_xCuO_4$	38
$YBa_2Cu_3O_7$	92
$Tl_2Ca_2Cu_3O_{10}$	125

instantaneous position of the electron. The ions are constrained by the restoring forces acting on them to vibrate about their positions of equilibrium and so the electron sets up an oscillatory positive charge in its wake. A second electron with the right momentum and phase is attracted to this region of excess positive charge to lower its own potential energy. If the two electrons are to interact over a substantial period of time, the two electrons must travel along the same line; it turns out that the electron pairs that do so are those of opposite momenta and of opposite spin. Other electron pairs can act similarly to enhance the effect and in a cooperative way lower their energy. The electrons involved must thus move in step in a highly organised manner to maintain this reduction in energy. We assume in what follows that the properties of the *lattice* are largely unchanged by the behaviour of the electrons and it is the *electrons alone* that determine any change in the free energy of the metal.

Consider the difference in free energy $\Delta F = \Delta E - T\Delta S$ between the ordered electrons and the randomly directed, normal electrons. If this becomes negative, the superconducting state has a lower free energy than the normal state and thus becomes the stable state. We have seen that the transition to superconductivity involves a reduction in the internal energy E (ΔE negative) but it also involves an ordering process which therefore reduces the entropy S (ΔS also negative). The $T\Delta S$ term thus acts to raise the free energy of the superconducting state and there is a competition between the ΔE and the $T\Delta S$ terms. The $T\Delta S$ term tends to diminish and ΔE to increase as T is reduced and so at low enough temperatures ΔE wins: the superconducting state becomes the thermodynamically stable state.

The thermoelectric power of a metal measures the entropy transported by the conduction electrons. It vanishes in a superconductor, which demonstrates that the superconducting electrons have zero entropy. What happens to the entropy of the conduction electrons in a superconductor is thus: at T_c a fraction of them become ordered and lose all their entropy. As the temperature is lowered further, this fraction increases until all of them have zero entropy at absolute zero.

If we start at a very low temperature with all the electrons in the ordered superconducting state, it is clear that to excite an electron out of this cooperative state we must break up a Cooper pair and this will require substantial energy. Call this ε. When we raise the temperature of the metal the probability of finding such energy will vary with a Boltzmann-type factor as $\exp(-\varepsilon/kT)$ and so, because of the dominance of the exponential, the heat capacity and entropy will also vary in this way. At higher temperatures the temperature dependence will alter as more and more electrons return to the normal state.

The reasons for the vanishing of the electrical resistance is linked to the cooperative nature of the superconducting state. In the normal state individual electrons are easily scattered (in classical language *deflected*) by impurities and lattice vibrations and so produce resistance. In the superconducting state the cooperating electrons form a large coherent unit, which moves bodily through the metal to carry the current. This unit cannot be scattered unless a major part of it is disrupted. The coherent unit can adjust to macroscopic changes of direction in the conducting wire but is not disrupted by single impurities or perturbations on the atomic scale; moreover, the superconducting electrons short-circuit all the rest and so no resistance is seen.

Liquid He^3

If He were a normal substance it would, when cooled, condense to form a solid at its triple point. In fact because of its small atomic mass, the condensed phase has a large zero-point vibrational energy (see page 106). Moreover, because the crystalline form is considerably denser than the liquid its zero point energy tends to be higher and so the crystal does not form at pressures at which the liquid and vapour are in equilibrium. It requires a very much higher pressure to force the liquid to crystallise and so there is no normal triple point. Figure 15 shows schematically the general shape of the phase diagram of both isotopes of helium at low temperatures and it shows that the liquid can persist down to the lowest temperatures.

How does this liquid behave? He^3 atoms have a resultant spin 1/2 and obey Fermi–Dirac statistics. Over a wide range of temperatures the liquid behaves rather like a weakly interacting Fermi–Dirac gas. The reason for this weak interaction is that in a degenerate Fermi–Dirac fluid the possibilities of interaction are severely restricted by the Pauli exclusion principle; this follows from an argument similar to the discussion on page 134 leading to equation

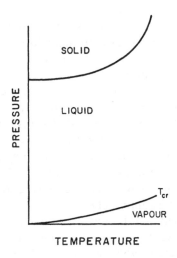

Figure 15 Schematic phase diagram of He^3 or He^4.

(163). Because, however, the He^3 atoms do interact with each other, they do not behave precisely as free particles but act as if the interaction enhanced their mass; the degeneracy temperature of the liquid is correspondingly lowered. At zero pressure, the effective mass of the liquid He^3 atoms is about twice the mass of the free atom and at high pressures near to the crystallisation pressure almost five times as big. The entropy of the liquid thus behaves rather like that of an ideal gas with an enhanced mass, with, of course, an important contribution from the nuclear spins, as we discuss below in the section on the melting curve.

The behaviour of the liquid as a quasi-ideal Fermi–Dirac gas, does not, however, persist down to the lowest temperatures. At about $1\,\text{mK}$ $(10^{-3}\,\text{K})$ the liquid undergoes a phase change to a superfluid phase, that is, a liquid whose normal viscosity vanishes and which can flow through channels without suffering any resistance. This is in some ways analogous to superconductivity but with uncharged particles. This phase, the A-phase, exists over only a small temperature range and is highly anisotropic; at still lower temperatures it gives way, with a small latent heat, to a second superfluid phase, the B-phase, which is almost isotropic in its properties. Both these transition temperatures depend on pressure and on magnetic field. The phase diagram of liquid He^3 in this region, in zero magnetic field, is shown schematically in Figure 16.

The explanation of this transition to superfluidity has some analogy with the BCS theory of superconductivity but with important differences. The mode of interaction in this liquid of neutral atoms cannot be through the Coulomb force as with electrons: it comes about through the nuclear spin which gives to each atom a small magnetic moment and gives rise to so-called exchange forces. As an He^3 atom travels through the liquid its spin causes interaction with the spins of nearby atoms and tends to make them point in the opposite sense to that of its own (antiferromagnetic coupling). It thus leaves a wake of partially polarised spins so that a second atom of the same spin as the first is able to use this environment to lower its own energy. In this way magnetic polarisation leads to an attraction between pairs of He^3 atoms and this interaction multiplied up by the cooperative effect of other pairs leads to a coherent long range interaction of many atoms, similar to that of electrons in a superconductor. The interaction via the spins, described above, leads to pairing of atoms with parallel spins and so to pairs with a total spin $S = 1$, rather than the anti-parallel spins of electron pairs in BCS superconductors. The pairs also have a resultant angular momentum with $L = 1$, which is often referred to as p-wave pairing. As in superconductors, a fraction of the material in this way acquires zero entropy and this fraction grows as the absolute zero is approached.

It must also be mentioned that there is a non-magnetic mode of interaction between He^3 atoms that leads to a mutual attraction and contributes to the superfluid state. Nonetheless what we have already seen of the spin-dependent interaction in liquid He^3 explains why the transition temperatures

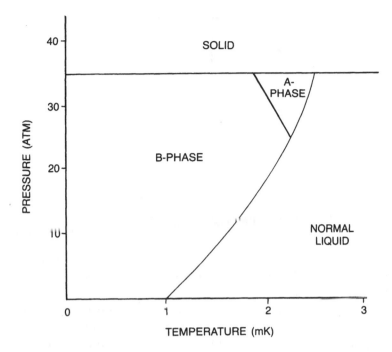

Figure 16 Approximate phase diagram of liquid He^3 in the mK range of temperatures to show the superfluid phases in zero magnetic field.

and properties of the superfluid depend on the applied magnetic field. In addition to this, a small magnetic field produces a third phase, called A_1, which exists only in the presence of such a field.

It is clear from this brief outline how subtle and varied are the ordering processes whereby the entropy tends to zero as absolute zero is approached. We turn now to a further example, liquid He^4.

Liquid He^4

Like liquid He^3 and for the same reason, liquid He^4 persists down to the lowest temperatures and it too has unusual properties. At a certain temperature, depending on the pressure, the heat capacity of the liquid behaves in a very strange way; it is illustrated in Figure 17. The shape of this curve resembles the Greek letter λ and the temperature at which the heat capacity changes abruptly is called the λ-point. Below this temperature the liquid becomes a superfluid.

This large anomaly in the heat capacity of the liquid is associated with a rapid decrease of its entropy with falling temperature (see Figure 18) and so we may think of it as due to some sort of ordering process. It is a

151

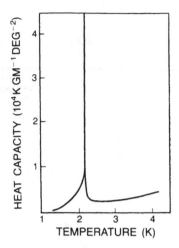

Figure 17 The λ-point in liquid He⁴. The curve shows the heat capacity of the liquid (under its saturated vapour pressure) as a function of temperature.

cooperative interaction involving many atoms and so cannot be treated by the statistical mechanics that we have used which apply only to weakly interacting atoms or molecules. However, the λ-phenomenon *is* related to the behaviour of a simple Bose–Einstein gas. Such a gas, as we have already seen, undergoes a 'condensation' at a certain critical temperature T_c depending on the density of the gas. An assembly of He⁴ atoms obeys Bose–Einstein statistics and although at the density of liquid helium there must be strong interactions between the atoms, it is of interest to compare the actual liquid with an ideal Bose–Einstein gas of the same density. The λ-point of liquid He⁴ in equilibrium with its vapour occurs at 2.19 K whereas T_c for a Bose–Einstein gas of He⁴ atoms at the same density is 3.14 K, which is not very different.

There is a stronger reason for thinking that the λ-point is associated with the Bose–Einstein statistics and that is that liquid He³, as we have seen, shows nothing comparable to the λ-phenomenon and certainly no superfluidity in the comparable temperature range.

Incidentally, part of the reason why we have made more progress in understanding superfluidity in He³ than in He⁴ (even though this was discovered almost fifty years earlier) is that, because of the Pauli principle that acts to reduce markedly the possible interactions between its atoms, He³ atoms can be treated as weakly interacting at densities and temperatures where He⁴ atoms cannot.

Mixtures

Consider a mixture of two isotopes or of two different elements and assume that it remains in true thermodynamic equilibrium down to the lowest temperatures. If the mixture is to obey the third law its entropy must tend to zero with the temperature, i.e. it must form some sort of ordered arrangement as T tends to zero. This can happen in several possible ways:

(i) On cooling the mixture, the mixture may separate into two distinct phases, one isotope or element being predominantly concentrated in one phase and the other isotope concentrated in the other. On cooling to the absolute zero, we might then expect to find two separate phases, each completely ordered. Phase separation has, for example, been observed in solid He^3–He^4 mixtures and is found in metallic alloys, although the process of separation tends to become very slow as the temperature is lowered.

(ii) In a mixture of two liquids it is possible for the contributions to the entropy to vanish without phase separation. In liquid He^3–He^4 mixtures, phase separation occurs over most of the concentration range but in mixtures containing less than a few percent of He^3 no phase separation occurs. These properties are important for the He^3–He^4 dilution refrigerator (see Chapter 15).

(iii) In a crystalline solid, the mixture can form an ordered arrangement in the lattice, a process that has been observed in alloys. Consider the simplest case in which there are equal numbers of two kinds of atom, A and B, say. At low temperatures an ordered array is formed in which A atoms and B atoms alternate in the lattice. At higher temperatures a disordering process begins and at very high temperatures the two kinds of atom are randomly distributed on the lattice sites. This process is a cooperative one (i.e. the behaviour of one atom influences and is influenced by that of the others) and is referred to as an 'order–disorder' transition.

The entropy associated with the fully disordered alloy, needed in the calculation of its free energy, can be found as follows. Consider N atoms of the alloy, made up of n_1 atoms of metal A and n_2 atoms of metal B with $n_1 + n_2 = N$. The two kinds of atom are distributed at random over the lattice sites and we wish to write down the number of distinguishable ways P in which this can be done. If all the atoms were different P would be $N!$. In fact, however, there are n_1 identical A atoms and n_2 identical B atoms. Therefore to each different arrangement of B atoms there are $n_1!$ arrangements of the A atoms that are indistinguishable. Likewise to each different arrangement of A atoms there are $n_2!$ indistinguishable arrangements of B atoms. Finally therefore:

$$P = \frac{N!}{n_1! n_2!}$$

So, since $S = k \ln P$, we find from Stirling's approximation:

$$\frac{S}{k} = N \ln N - n_1 \ln n_1 - n_2 \ln n_2$$

(having cancelled the linear terms). Putting $N = n_1 + n_2$ in the coefficient of $\ln N$, we can write this as:

$$\frac{S}{k} = -n_1 \ln \frac{n_1}{N} - n_2 \ln \frac{n_2}{N}$$

or for 1 mole of alloy ($N = N_0$, Avogadro's number, and $N_0 k = R$):

$$\frac{S}{R} = -x_1 \ln x_1 - x_2 \ln x_2$$

where x_1 and x_2 are molar concentrations. This result can readily be generalised to any number of components.

The fully disordered arrangement may persist down to the lowest temperatures because diffusion becomes so slow that the atoms cannot achieve their true equilibrium configuration. This true equilibrium configuration, if it could be reached, would consist of an ordered alloy or, depending on the composition, a mixture of ordered alloys (including the special case of pure components). These would satisfy the third law and have zero entropy. We discuss these non-equilibrium states more fully below.

Entropy differences

We can look at the third law in a slightly different way by examining the difference in entropy of two states of the same system at the same temperature. If the path between them is reversible, the third law demands that these differences should tend to zero as the absolute zero is approached. Here are some examples.

The thermal expansion coefficient

The (volumetric) thermal expansion coefficient is defined as $(1/V)(\partial V/\partial T)_p$. But

$$\left(\frac{\partial V}{\partial T}\right)_p = -\left(\frac{\partial S}{\partial p}\right)_T \tag{166}$$

by a Maxwell transformation. Now the right hand side of this equation measures the rate of change of entropy with pressure at constant temperature. If the entropy difference due to a small isothermal pressure difference is to tend to zero as $T \Rightarrow 0$, then $(\partial S/\partial p)_T$ must tend to zero. Thus the coefficient $(\partial V/\partial T)_p$ must likewise vanish as $T \Rightarrow 0$, and so must the thermal expansion coefficient. One deduction from the third law is therefore that the expansion coefficients of solids and liquids must vanish at the lowest temperature. This is confirmed by experiment.

The magnetic susceptibility

If M is the magnetic moment of a substance in a magnetic field B at temperature T, a Maxwell transformation shows that

$$\left(\frac{\partial M}{\partial T}\right)_B = \left(\frac{\partial S}{\partial B}\right)_T \tag{167}$$

By the third law of thermodynamics $(\partial S/\partial B)_T$ must vanish as $T \Rightarrow 0$. Thus $(\partial M/\partial T)_B$ must also vanish.

The magnetic susceptibility of many paramagnetic insulators obeys the Curie law. The magnetic susceptibility χ is defined by the relation $M = \chi H$ where $H = B/\mu\mu_0$. In the weak paramagnetic materials with which we deal μ can be taken as unity; μ_0 is defined, in the M.K.S. system, as $4\pi \times 10^{-7}$ (newton/ampere^2). For a material that obeys the Curie law, $\chi = c/T$ where c is a constant. For these substances therefore:

$$\left(\frac{\partial M}{\partial T}\right)_B = -\frac{cB}{\mu_0 T^2} \tag{168}$$

and as $T \Rightarrow 0$, this, far from going towards zero, tends to infinity. Thus the third law implies that at low enough temperatures the Curie law must fail. Physically this means that the magnetic dipoles which give rise to the magnetic moment M must interact with each other and at low enough temperatures this interaction must alter the magnetic properties of the substance. This is indeed found experimentally. The Curie law holds only at temperatures so large that the thermal energy of the elementary dipoles is much greater than their mutual interaction energy: at sufficiently low temperatures, some sort of ordering process is found to occur (e.g. a transition to a ferromagnetic or antiferromagnetic state).

The melting curve

According to the Clausius–Clapeyron equation, the slope of the melting curve is related to the entropy change ΔS and the volume change ΔV on melting by the equation:

$$\frac{\mathrm{d}p}{\mathrm{d}T} = \frac{\Delta S}{\Delta V} \tag{169}$$

At constant pressure, melting is an isothermal process. According to the third law the entropy difference between the solid and liquid form of a substance at the same temperature must tend to zero as $T \Rightarrow 0$. As we have seen, both He^4 and He^3 form liquids that exist down to the lowest temperatures (cf. Figure 15). Under suitable pressures the liquid and solid exist together at these low temperatures and the behaviour of ΔS can be studied. According to the third law ΔS must vanish as $T \Rightarrow 0$ so that, unless ΔV also vanishes (this is not so for either isotope), the slope of the melting curve dp/dT must also vanish. Let us compare what happens in the two helium isotopes.

He^4

Solid He^4 behaves more or less normally at low temperatures. On the other hand, as we have seen, the liquid has unusual properties. Figure 18a shows how the entropy of the solid in equilibrium with the liquid and the liquid in equilibrium with the solid go towards zero as the temperature tends towards absolute zero. The melting curve is shown in Figure 18b and you can see that it quickly becomes flat below the λ-point. There is in fact a small minimum in this curve around 0.7 K but there is every reason to suppose that it remains essentially flat down to the absolute zero.

At temperatures below about 1 K, the solid and liquid have such small entropies that there is almost no latent heat of solidification and all the energy involved in the process is in the form of the work done in reducing the volume from that of the liquid to that of the solid. The relative stability of solid and liquid is here determined simply by the mechanical criterion: which has the lower energy? In this region the free energy is essentially the same as the internal energy and the process of freezing and melting is a mechanical one.

He^3

Liquid He^3 is so different from liquid He^4 that the melting curve of He^3 is bound to differ substantially from that of He^4. However, the solid He^3 too plays a quite different role because the nuclear spin gives it an additional entropy of $R \ln 2$ (the nuclear spin of He^3 is 1/2), which remains until magnetic ordering of the nuclear spins begins at around 10 mK. Below 1 K the entropy from the lattice vibrations is small enough to be ignored.

The entropy of the liquid at temperatures above the superfluid transition is similar, as we have seen, to that of Fermi–Dirac gas with a characteristic temperature T_F of about 1 K (five times smaller than that of an ideal gas of He^3 at the same density). The entropy here consists of a component due to the motion of the atoms and a component from the spins.

The entropies of solid and liquid in mutual equilibrium are shown schematically in Figure 19(a). From this it is clear that the entropy of the liquid equals that of the solid at about 0.3 K and becomes smaller at lower temperatures. Thus the entropy difference $\Delta S = S_{\text{liquid}} - S_{\text{solid}}$ is positive above 0.3 K and negative below it. Since ΔV, the volume difference between liquid and solid, remains roughly constant and positive throughout, it follows from the Clausius–Clapeyron equation:

$$\frac{dp}{dT} = \frac{\Delta S}{\Delta V}$$

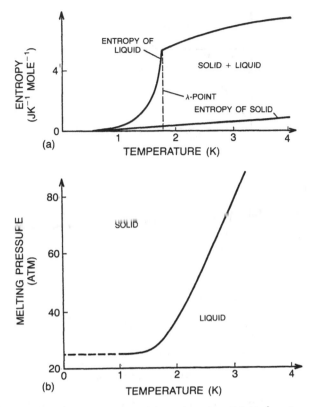

Figure 18 (a) The entropy diagram of liquid and solid He^4 at low temperatures. (b) The corresponding part of the melting curve of He^4.

that the slope of the melting curve must be positive at high temperatures and become negative below 0.3 K i.e. the melting curve must go through a minimum. This is indeed found (Figure 19(b)). Consequently, if you start with liquid He^3 at the pressure and temperature corresponding to the point A, say, and warm it at constant pressure, it first solidifies (at B) and then melts again (at C). The idea that you can add heat to a liquid and cause it to freeze is a strange one. Usually you add heat to a solid to melt it because the liquid has a higher entropy than the corresponding solid. Generally this is true because the molecules in the liquid are spatially more disordered than in the solid. However, in He^3 at these low temperatures the nuclear spin entropy is decisive and, because the nuclear spins are more disordered in the solid than in the liquid, the solid here can have the greater entropy.

In the case of He^3, the melting curve, at least in this temperature range, does not of itself provide a test of the third law but does bring to light this unexpected behaviour of the entropy.

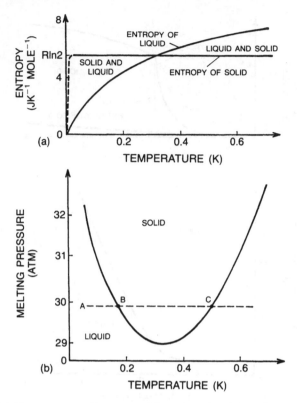

Figure 19 (a) The entropy diagram of liquid and solid He³ below 1 K. (b) The corresponding part of the melting curve of He³.

Heat capacities near the absolute zero

Consider first a system at fixed volume; the difference between the entropy of the system at temperature T and its value at $0\,$K is given by

$$S_T - S_0 = \int_0^T \frac{C_V}{T}\,\mathrm{d}T$$

where C_V is the heat capacity at constant volume. According to the third law the constant S_0 is zero; as I stressed earlier, any finite value for S_0 would do, but since it is to be the same for all systems and since its value is at our disposal zero is the obvious choice. Therefore to satisfy the third law the integral on the right-hand side of the equation must be finite. (This must be true if S_0 is to have any finite value.) If, however, C_V remained constant down to the absolute zero, this integral would diverge. Indeed, if the integral is to remain finite the specific heat must tend to zero as $T \to 0$. We therefore deduce from the third law that C_V for all substances must tend to zero at the absolute zero. A similar argument applies to C_p, the specific heat at constant pressure.

We have already seen in our statistical treatment of the heat capacities of some simple models of solids and of the ideal Bose–Einstein and Fermi–Dirac gases that C_V in all these examples does indeed tend to zero as $T \to 0$. This is a result of quantum mechanics. Because the energy levels of atomic systems are quantised, it is always possible to find a temperature so low that kT becomes comparable with the separation between the lowest-lying energy levels. When this happens the specific heat must fall off and at the lowest temperatures when all the elements in an assembly are effectively in their lowest energy state the specific heat must vanish.

The vanishing of the heat capacity therefore, although a consequence of the third law of thermodynamics in the form that we have stated it, is most readily understood in terms of the quantum mechanical behaviour of atoms.

Gases and the third law

In discussing the statistical mechanics of gases we showed that the entropy of both an ideal Bose–Einstein gas and an ideal Fermi–Dirac gas vanishes at the absolute zero, provided only that the molar volume of the gas remains finite. These ideal gases therefore obey the third law of thermodynamics.

Moreover, there are some examples of actual systems which behave like gases at low temperatures and whose properties can be studied down to the lowest temperatures. One such example is the electron gas in a metal. Because the contribution to the heat capacity from the lattice vibrations falls off as T^3 at sufficiently low temperatures, the contribution from the electrons in a metal predominates and can readily be measured experimentally. If the metal does not become super-conducting, the heat capacity (and hence the entropy) is found to vary linearly with T like that of an ideal Fermi–Dirac gas. In a superconductor the heat capacity and the entropy of the electrons fall off to zero much more rapidly than this below the superconducting transition temperature, approximately as $e^{-\eta/kT}$, where η is some characteristic energy. Thus we see that these 'gases' also obey the third law of thermodynamics.

If we take an actual gas (one we can put into a gas jar) and cool it at, say, constant pressure we find that ultimately it turns into either a solid or a liquid. Thus if we wish to study the properties of the gas (or more strictly the vapour as it now is) at lower temperatures its pressure must at each temperature be kept lower than or equal to the equilibrium vapour pressure at that temperature; otherwise it will condense. So on going down in temperature the entropy of the vapour changes both because the temperature changes and because the vapour pressure changes. The lower temperature tends to *reduce* the entropy but the lower vapour pressure tends to *increase*

159

it. Moreover the vapour pressure goes down so rapidly with temperature (roughly as $e^{-L/RT}$, where L is the latent heat of vaporisation) that there is a net *increase* in the entropy (per unit mass) of the vapour. Thus its entropy, far from going towards zero, tends to get bigger as the temperature goes down (cf. Figure 20).

Is this then a violation of the third law of thermodynamics? The answer is no, for the following reason: at a low enough temperature (but still above the absolute zero) the vapour ceases to exist. For example, the vapour pressure of He^4 at 0.03 K is estimated to be 6×10^{-103} mm of mercury so that to enclose one *single molecule* of the vapour at this temperature would require a volume approximately equal to that of the universe. In other words, at this temperature there is no He^4 vapour. This is equally true of He^3 at low enough temperatures and even more true of all other substances; consequently it is impossible to find the limiting value of the entropy of a vapour as T tends to zero, because in this limit *there are no vapours*.

Because all real gases cease to exist at very low temperatures we cannot *directly* test the predictions of the third law on them. Nevertheless the third law can be applied to real gases in the following way. The translational entropy of a real gas can be calculated at some high temperature and low density by making use of the Sackur–Tetrode equation (equation (144) Part 2) for the entropy of an ideal classical gas. At sufficiently high temperatures and low densities the corrections to be applied to this because the actual gas is not ideal are small and can be deduced from experimental data. Moreover, the contributions to the entropy of the gas from molecular vibrations and rotations, from electronic excitations and other additional sources of entropy, can be worked out from a detailed study of the optical and other spectra of the gas. In this way, the total entropy of the gas can be computed accurately using the methods of statistical mechanics. The value of the entropy computed in this way may be called the statistical entropy of the gas, S_{stat}.

The entropy of the gas can be found in a different way; its value relative to the absolute zero can be determined directly by heat capacity measurements down to very low temperatures. To find the entropy in this way, we measure the heat capacity of the *gas* at constant pressure down to the temperature at which it condenses to a liquid. We measure the entropy of condensation from the latent heat and we then measure the heat capacity of the *liquid* until it reaches the solidification temperature. We now measure the entropy change on solidification, again from the latent heat, and then we measure the heat capacity of the *solid* down to the lowest temperatures. If we make measurements to a sufficiently low temperature we can then extrapolate to the absolute zero and so, by integration of the heat capacity divided by the absolute temperature, obtain the entropy of the gas at the temperature at which we started. The value of the entropy obtained in this way is called the 'calorimetric' entropy, S_{cal}. An example of the various contributions to S_{cal} is given in Table 9.

Clearly the calorimetric entropy (if it has been determined correctly) measures the difference in entropy between the solid at 0 K and the gas under its initial conditions (the conditions of interest to us).

On the other hand, we have already seen that the Sackur–Tetrode equation, in the domain where it is valid, measures the translational entropy of a gas relative to its value at 0 K. This is also true of the vibrational, rotational and electronic contributions to the entropy of the gas. Therefore the total statistical entropy of the gas measures the difference in entropy between the gas at 0 K and the gas under the conditions in which we are interested.

Now the third law comes in. According to this the entropy difference between *all* equilibrium states of a system vanishes at 0 K. Therefore we can conclude that the solid at 0 K must have the same entropy as the gas at 0 K (if it could exist there). Therefore since S_{stat} and S_{cal} are both measured with respect to the same origin, and both refer to the same final conditions, they should be equal. That is, if the third law holds.

To check that the third law is valid we can thus compare the entropies deduced in these two ways and see if they are the same. Usually the contributions of the nuclear spins are ignored on the assumption that they will be the same in the solid (at the lowest temperatures normally measured) as in the gas. In Table 10 some comparisons of this kind are made and these show that S_{stat} and S_{cal} do agree well, but this is not always so.

Now that the third law is accepted as valid, a disagreement between S_{stat} and S_{cal} is taken to mean that something is wrong with our estimates of one or both. For example, we may have neglected some contribution to the statistical entropy or our calorimetric determination may be unsatisfactory. The latter can occur either if our extrapolation to 0 K from the lowest temperature of measurement neglected an important contribution to the entropy or if at some stage of the cooling our system ceased to be in internal

Table 9 Calorimetric entropy of argon gas at its normal boiling point. Contributions from various temperature ranges

0–10 K (extrapolated)	0.303 cal K^{-1} mole
10–83.78 K solid	8.815
Entropy of melting at 83.78 K	3.351
83.78 to 87.29 K liquid	0.413
Entropy of vaporisation at 87.29 K	17.843
Total entropy of gas at one atmosphere pressure at 87.29 K	30.725
Correction for non-ideality of gas	0.125
Calorimetric entropy of argon at its normal boiling point in the (hypothetical) ideal gas state	30.85 ± 0.10

(These data are taken from the work of Clusius and Frank[2] in 1943)

thermodynamic equilibrium. As we shall see below, when this happens we can no longer calculate the entropy from the apparent heat capacity.

Non-equilibrium states

So far we have assumed that our systems could undergo reversible changes down to the lowest temperatures. However, this is not always, or even usually, the case in practice. In discussing mixtures of isotopes I quoted the example of He^3–He^4 mixtures which even in the solid phase show phase separation at low enough temperatures. With most isotopic mixtures, however, this sort of separation cannot occur because, at the temperatures at which, from a purely thermodynamic point of view, the phase separation should begin, diffusion is so slow that the process virtually cannot take place at all. Consequently, most solids at low temperatures consist of disordered mixtures of isotopes which are therefore strictly not in thermodynamic equilibrium. The positive assertions of the third law cannot, therefore, be applied to the entropy associated with the isotopic mixing. The important point is, however, that the third law can still be applied to other entropy changes that the solid can undergo reversibly at low temperatures. For example, its thermal expansion coefficient must still vanish as the third law requires; if it is a paramagnetic solid, obeying Curie's law at higher temperatures, this law must ultimately fail at some low temperature and some kind of ordering process of the atomic magnets must take place.

Another example of frozen-in disorder is provided by glasses. If a liquid is cooled below its normal freezing point it may remain in a metastable state and be effectively in internal thermodynamic equilibrium. With certain liquids, for example glycerol, it is possible to reach a temperature at which diffusion in the liquid becomes so slow that the molecules can no longer assume their positions of true thermodynamic equilibrium (corresponding to the metastable liquid phase) and the phase behaves more like a solid than a liquid. The disorder in the atomic positions appropriate to the liquid at a certain temperature is frozen in and at lower temperatures remains unchanged. This is also true of some very rapidly quenched metallic alloys.

Table 10 Some examples of calorimetric and statistical entropies

	Calorimetric entropy	Statistical* entropy
Argon	30.85	30.87
Oxygen	40.74	40.68
Nitrogen	36.53	36.42

All the gases are compared at their normal boiling points
*These values omit any contribution from nuclear spins

In this condition the substance is referred to as a glass and because it is no longer in internal equilibrium we cannot apply the positive assertions of the third law to this aspect of it. For example, we cannot make any definite statement about the configurational entropy differences between the glass and the same material in its crystalline form. This is perhaps not surprising since we can no longer measure the entropy difference between the glass and the crystal because no reversible path exists between them. In spite of this, the third law can still be applied, for example, to the changes of entropy of the glass with volume and we may expect its thermal expansion coefficient to vanish as T tends to zero.

The limitations that we have imposed on the applications of the third law in its positive form, namely that it shall apply only to systems or parts of a system in internal thermodynamic equilibrium, are, however, restrictions that must be imposed on the first and second laws when stated in their positive forms. Entropy differences cannot be defined unless there exists a reversible path between the states under consideration and changes in the internal energy function can also only be defined if the initial and final states are truly in equilibrium.

The third law and chemical equilibrium

The examples of the application of the third law that we have looked at so far have been mainly concerned with the entropy of different physical states of the same system. However, the third law allows us to compare the entropy of systems in different states of *chemical* combination, e.g. the entropy of half a gram molecule of chlorine and one gram molecule of sodium, on the one hand, with the entropy of a gram molecule of NaCl on the other. This sort of comparison is important in the discussion of chemical equilibrium. When a group of reacting substances are in chemical equilibrium the free energy change associated with the reaction is then zero. To determine the condition of chemical equilibrium *a priori* therefore involves knowing the free energy of the reactants and this in turn demands a knowledge of their entropies. This was in fact the original purpose for which the Nernst heat theorem was formulated. By means of the heat theorem chemical equilibrium constants could be obtained from purely thermal measurements. Today, however, equilibrium constants are usually calculated directly by means of statistical mechanics on the basis of data obtained from atomic and molecular spectroscopy.

The third law – a summary

A convenient and general statement of the third law was given by Simon in 1937. It is as follows[3]: '*The contribution to the entropy of a system from each sub-system which is in internal thermodynamic equilibrium disappears at the absolute zero.'*

On the basis of the statistical interpretation of entropy, this means that systems or parts of a system in thermodynamic equilibrium tend to a state of order at the absolute zero. Examples of order, in this sense, are the spatial order of atoms in a perfect lattice or the regular arrangement of different kinds of atoms in the lattice of an ordered alloy. In some solid and liquid mixtures, the ordering is produced by the complete separation of the components from the mixture. There is also the directional ordering of the atomic magnets in the ferromagnetic or antiferromagnetic state.

In general it is the mutual interaction of the atoms or molecules in an assembly which is ultimately responsible for bringing about these states of order. When kT becomes small compared with this interaction energy (per atom or molecule) the ordering mechanism can be expected to show itself. On the other hand, although statistical mechanics can re-express the significance of the third law it cannot, as things stand at present, prove it. The third law still remains a generalisation, based on experimental observation, about the behaviour of the thermodynamic properties of matter at low temperatures.

With this summary we will leave the third law of thermodynamics and turn to the methods of measuring absolute temperatures especially at low temperatures. After that, we will have a brief look at a slightly different version of the law.

Absolute temperature and low temperatures

The definition of entropy shows that the concepts of entropy and absolute temperature are inseparably linked so that, if we are to measure entropy, we must first be able to measure absolute temperature.

Measurement of absolute temperature

We have seen already that the Kelvin absolute temperature was defined to be identical with the ideal gas scale and so, as long as we have real gases whose equation of state can be accurately measured at very low pressures, we have a practical way of realising the absolute scale of temperature. The practicality of this, however, disappears at both low and high temperatures: at high temperatures because it becomes impossible to contain the gas (the containers soften or melt); at low temperatures because gases effectively cease to exist.

Measurement of high temperatures

At high temperatures we use the properties of *black body* or *cavity* radiation; such properties are independent of the material that makes up the cavity and depend only on temperature and the frequency range studied. It is worth recalling that it was Planck's efforts to predict the entropy of black body radiation that led to the full understanding of this radiation and also to the formulation of quantum theory. It is the Planck radiation formula which enables us to use cavity radiation to measure absolute temperatures above those accessible to a gas thermometer; it has a powerful theoretical foundation and in the realm where it can be tested experimentally by the use of gas thermometry it has been fully tested and verified.

The use of cavity radiation to measure temperature both on earth and in astronomical objects is a vast subject and quite beyond the scope of this book. The important feature is that it does provide a practical way of determining absolute temperatures in this otherwise almost inaccessible region and has had many remarkable successes.

Low temperatures

The third law of thermodynamics shows that the natural zero of entropy is at the absolute zero of temperature and so entropy is also inevitably linked to the study of its behaviour at low temperatures. Moreover the methods of measuring absolute temperatures near absolute zero are closely linked to the methods of producing them. We therefore digress briefly to look at methods of producing low temperatures before we consider how to determine their absolute values.

How low temperatures are produced

When we discussed heat engines (Chapter 4) we saw that a reversible heat engine can be used as a refrigerator: indeed it is clear that such an engine working backwards is the most efficient refrigerator possible. When it is so used its lower operating temperature is generally variable while its upper operating temperature is either room temperature or that of some convenient low temperature bath. Although the efficiency of the machine is, in principle, independent of the working substance, in practice the choice of working substance is very important. This is partly because we can never achieve the ideal reversibility of the Carnot engine, partly because the rate of cooling becomes of importance in real low temperature machines and partly because we must have a practical and convenient way of abstracting the heat at the high temperature.

In practice therefore we look for two important characteristics in the working substance if we are to use something approaching a Carnot cycle in the actual operation of our machine:

(i) The working substance should have a large entropy which can be reduced by straightforward practical methods at the lowest convenient starting temperature.

(ii) The substance should be capable of reversible (or nearly reversible) changes down to the lowest temperatures at which it is to be used. Down to about 1 K, gases can be found that satisfy both these requirements. As working substances, gases are convenient; they can easily be compressed and then sent through pipes to where they are needed. Moreover, liquefied gases with their comparatively large latent heats form convenient constant temperature baths. Thus until the 1920s the liquefaction of gases was the dominant method of producing low temperatures. Since then the magnetic method of cooling and more recently the use of He^3–He^4 dilution refrigerators have come into prominence. We briefly look at these three methods.

The liquefaction of gases

There are two main methods for liquefying gases, both in commercial use. These are (1) Joule–Thomson cooling and (2) the external work method. Both these methods make use of the heat exchanger, a device whereby the cold gas leaving the liquefier cools the incoming gas so that the cooling becomes cumulative and finally leads to liquefaction. Usually these heat exchangers consist of two long tubes joined together side by side in good thermal contact with each other but otherwise thermally insulated; they are then wound into a spiral. The compressed gas goes down one tube and then returns after cooling via the other; in this way the outcoming gas cools that going in. If the heat exchanger is perfect, the outgoing gas leaves at exactly the same temperature as the gas entering it.

(1) Joule–Thomson cooling

This method was devised by Linde (1842–1934) and is often called the Linde method. It is based on the Joule–Thomson effect, which we have already described. We saw that if a gas is forced under pressure through a constriction there is a change in its temperature and in certain pressure ranges there is a cooling effect. This occurs at temperatures below a certain temperature, known as the inversion temperature, and clearly the gas must be cooled below this before it can be further cooled by a Joule–Thomson expansion. The Joule–Thomson effect is essentially irreversible and so a liquefier based on it starts at a disadvantage. Nevertheless, since there are no moving parts of the liquefier at low temperatures, the Linde method is mechanically so simple and direct that these advantages largely make up for the intrinsic inefficiency.

Some details of the method are brought out in the examples relating to this chapter.

(2) The external work method of liquefaction

The adiabatic expansion of a gas is an obvious way of producing low temperatures and was exploited by Claude and Heylandt for the liquefaction of air. In these machines the air is compressed at room temperature and allowed to expand adiabatically doing work in a continuously operating engine. Liquefaction in the actual cylinder of the engine leads to erratic, even explosive, operation and so the final stage of liquefaction is achieved by a Joule–Thomson expansion.

The same principle of external work under adiabatic (or nearly adiabatic) conditions is used in the Collins helium liquefier. This commercial liquefier has made the production of liquid helium, once the prerogative of only a few specialised laboratories, easy and commonplace in laboratories throughout the world. In this liquefier, too, the actual final liquefaction is made by means of a Joule–Thomson expansion. The advantage of the external work method, apart from its intrinsic efficiency, is that it can cool the helium gas below its inversion temperature.

A different type of helium liquefier using a single stroke adiabatic expansion process is the Simon expansion liquefier, which, however, is now of largely historical interest since it requires the use of liquid hydrogen for precooling the helium (see exercise 1 of this chapter).

Cooling by evaporation of liquefied gases

Once the liquid is produced it now provides a rather useful way of going to lower temperatures. To do this you reduce the pressure over the liquid under adiabatic conditions. The reduced pressure causes the liquid to evaporate; the evaporation removes the latent heat of vaporisation from the system and causes the temperature to fall.

The process can be followed more exactly on an entropy–temperature diagram. Suppose that we have now produced a liquid boiling under pressure. Let it be represented by the point A in the S–T diagram (Figure 20). T_t is the triple point temperature, BC is the entropy of vaporisation and CD the entropy of melting at this temperature.

When we slowly reduce the pressure over the liquid this is equivalent to a reversible expansion of the system. Since the system is now adiabatic, its representative point moves at constant entropy, i.e. along the line AE into the region of lower pressures. This takes it into the two-phase region where the vapour and liquid co-exist. The lowest temperature that can be attained in the liquid phase is the triple point temperature, T_t. At this temperature the system would have its representative point at E. This means (see p. 47) that some of the liquid has vaporised and the ratio of the mass of liquid to the mass of vapour is seen from the diagram to be BE/EC. If the liquid has to cool some piece of apparatus or another gas, in addition to itself, the line it follows on the S–T diagram will no longer be horizontal. It will rise by ΔS, where ΔS is the entropy entering from the other piece of apparatus. Thus the fraction of vapour produced will be bigger. By the same method, you can also go below the triple point temperature into the region where the solid

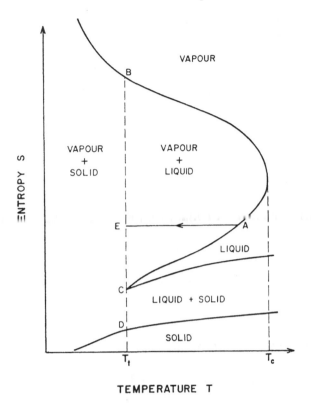

Figure 20 Entropy diagram of vapour, liquid and solid below the critical temperature, T_c. T_t is the triple point temperature. The process *AE* represents a reversible adiabatic expansion of the liquid *A* to form a mixture of vapour and liquid as at *E*.

and vapour are in equilibrium. However, the vapour pressure of the solid falls off rapidly as the temperature goes down and the method fairly soon becomes impractical. For example, with liquid hydrogen, whose boiling point is 20.4 K, the triple point at 14 K is easily reached by pumping (the vapour pressure is here about 5 cm Hg). The lowest temperature to which the solid can conveniently be cooled by pumping is about 10–11 K. There is a further practical point about *solidified* gases; they make rather poor low temperature baths because it is often difficult to make good thermal contact with the solid.

Because liquid helium cannot be solidified by pumping, it forms a very useful low temperature bath. In Table 11 are listed the vapour pressures of liquid He^4 and He^3 at a number of temperatures. You might expect from the table that liquid He^4 would be useful at least down to 0.5 K, but in fact it is not readily usable below about 1 K. This is because liquid He^4 has peculiar flow properties below the λ-point (2.2 K) and these cause excessive evaporation of the liquid when wide pumping tubes are used. Without any

169

Table 11 Vapour pressure of helium

T(K)	He^4 p (mm)	He^3 p (mm)
1.0	1.20×10^{-1}	8.56
0.7	2.27×10^{-3}	1.29
0.5	1.63×10^{-5}	1.4×10^{-1}
0.3	$\sim 10^{-10}$	1.5×10^{-3}
0.1	$\sim 10^{-31}$	
0.03	$\sim 10^{-103}$	

special pumps or oddities of design, temperatures down to 1.2 K can easily be reached by pumping on a liquid He^4 bath; however, to go much lower than this becomes complicated and even then 0.7 K is about the lowest temperature that can be reached in this way. Liquid He^3, on the other hand, having no exceptional flow properties at these temperatures, can easily be used down to about 0.3 K (see Table 11).

The He^3–He^4 dilution refrigerator

This refrigerator, originally proposed by H. London in 1951, exploits the comparatively large entropy of solutions of He^3 in liquid He^4. In such solutions there is an entropy of mixing which depends on concentration and temperature. Instead of thinking in terms of the entropy of mixing we can to a good approximation neglect the contribution of the He^4 and treat the He^3 atoms rather like a gas within the volume provided by the liquid He^4. Thus dilution of the solution is equivalent to an expansion of the quasi-gaseous He^3 and, as with a real gas, the adiabatic expansion causes cooling. Because the entropy of the liquid He^4 is negligible below about 1 K, the He^4 scarcely impedes this cooling.

The objective then is to start at the lowest temperature attainable by conventional methods with a high concentration of He^3 in liquid He^4 and to dilute it reversibly and adiabatically. As we shall soon see, the phase separation that occurs in liquid He^3–He^4 solutions provides a convenient way to achieve this dilution.

In practice the dilution refrigerator operates in a continuous flow mode in which gaseous He^3 is first cooled to 4.2 K by means of a bath of liquid He^4 at its normal boiling point and then condenses at 1 K, cooled by a bath of pumped He^4. It then passes through heat exchangers to the 'mixing chamber' in which there is a two-phase liquid mixture of He^3 and He^4, with the He^3 rich phase (almost 100% He^3) floating on the He^4 rich phase (about 6% He^3). When He^3 diffuses across the phase boundary cooling occurs through the dilution process outlined above. The cooled He^3 passes out

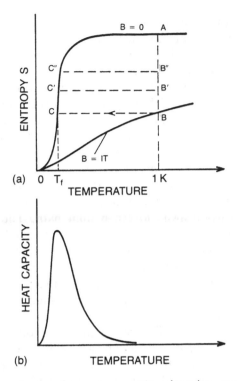

Figure 21 (a) The entropy of a paramagnetic salt at low temperatures in zero field and in a high magnetic field (cf. Figure 11). (b) The heat capacity of the salt in zero field.

through the conduit of liquid He^4 via the heat exchangers, thereby cooling the incoming liquid. It then reaches the so-called 'still', which is kept at about 0.6 K, where the He^3 evaporates much more strongly than the He^4 because of its much higher vapour pressure (see Table 11). The He^3 is then returned for recirculation.

In this way steady temperatures of about 10 mK are routinely achieved and even lower temperatures, down to about 3 mK, can be obtained with more refined laboratory refrigerators.

The magnetic method of cooling

The magnetic method of cooling has a long history: it was first suggested in about 1925 and first used in 1933. At that time when no He^3 was available, temperatures down to about 1 K could be achieved by pumping on liquid He^4; to go still lower a source of high entropy that could readily be reduced by straightforward, practical means and capable of reversible changes at the lowest temperature, was needed. Certain paramagnetic salts (for example,

171

iron ammonium sulphate, chromium potassium alum) have such properties. These solids contain ions (Fe, Cr) that carry a permanent magnetic moment and so behave like tiny atomic magnets; they can thus be the source of high entropy, as we discussed in Chapter 10.

To be useful for low temperature work, the ions must interact with each other so weakly that even at 1 K they are effectively independent of each other. As we have seen, this cannot persist down to the absolute zero because at some low temperature the mutual interaction between the magnets or between the magnets and the lattice must lead to an ordering process, for example, a transition to ferro- or antiferromagnetism.

The method consists of magnetising the salt at, say, 1 K; the heat of magnetisation, evolved when the entropy of the salt is reduced by the magnetic field, is carried away to the helium bath. The salt is thermally isolated with the field still on; the field is then reversibly reduced to zero and this causes the temperature of the salt to fall. An entropy–temperature diagram illustrates the process (Figure 21(a)). This shows the entropy of a typical salt below 1 K in zero field. At 1 K the entropy of the salt is entirely due to the magnetic dipoles in the salt; the contribution from lattice vibrations is negligible. In zero field at 1 K the dipoles are randomly oriented and have high entropy as shown at A. This randomness is reduced, and with it the entropy, by applying a magnetic field, typically about 1 tesla. The representative point of the substance thus moves from A to B. If now the salt is thermally isolated and the field is reduced slowly and reversibly to zero, the representative point moves at constant entropy from B to C on the B = 0 line, thereby reaching a low temperature. The precise value of this temperature obviously depends on the starting field and on the temperature at which magnetic ordering sets in.

Notice that the heat capacity of the salt is likely to be large at the final temperature because this occurs just where the entropy of the salt in zero field is falling fast. Its heat capacity $T(\partial S/\partial T)_B$ is determined by the slope of the $S(T)$ curve in zero field, which as illustrated in Figure 21(b) is large at these temperatures. Conversely around 1 K where the $S(T)$ curve is flat the heat capacity is tiny. To give some idea of the magnitudes involved let me quote an estimate that $1\,cm^3$ of iron ammonium alum at about 50 mK (the temperature where its heat capacity has a maximum) has a heat capacity equal to that of about 16 tons of lead at the same temperature. From the point of view of holding the low temperature this is very convenient!

The temperatures which can conveniently be reached by the adiabatic demagnetisation of paramagnetic salts are of the order of 10 mK. Today such temperatures are more readily achieved and held by means of the He^3–He^4 dilution refrigerator. Nonetheless the method can be adapted to produce very much lower temperatures by using, instead of the paramagnetism due to electrons, that of the nuclear spins.

The nuclear spins have magnetic moments which are smaller than those of electrons by about the ratio of the mass of the proton to that of the electron, i.e. about 2000. This and the kind of interaction between nuclei reduce the strength of mutual interaction, reduce the temperature at which ordering begins and lower the lowest achievable temperatures. That is, of course, provided that an appreciable fraction of the entropy present at the starting temperature can be removed. This is made harder by the smallness of the nuclear magnetic moment and so requires a low starting temperature and a high initial magnetic field. In Chapter 10 we saw that the important quantity is the ratio of starting field B to starting temperature T. Suppose we start at 10 mK with a field of 10 T, which is possible with modern magnets, giving $B/T = 1000$. This is to be compared with 1 T at 1 K in the earlier experiments using electron paramagnets but with an additional factor of 2000 to take account of the bigger magnetic moment. So modern techniques achieve about the same fractional reduction in entropy of the nuclear spins as earlier techniques achieved with electron paramagnets. There are more sophisticated methods of polarising nuclei using the coupling between electron and nuclear spins or between the lattice and the nuclear spins but these do not alter the principles involved.

There are, however, two practical problems to be overcome. One is to find a suitable heat switch that allows the nuclear spin material (usually copper) to be cooled to the starting temperature and to carry away the heat of magnetisation. It must then be capable of thermally isolating the copper from its surroundings for the adiabatic demagnetisation. The switch is usually a superconductor that exploits the poor thermal conductivity of the superconducting state as compared to the normal state induced by the magnetic field. The second problem is that, although it may be possible to cool the nuclear spin system, it becomes increasingly difficult at these low temperatures to couple the spins to the rest of the metal and thus to cool it and any attached specimen. The nuclear spin relaxation time, which is a measure of the time taken to move towards thermal equilibrium between the nuclear spins and the metal as a whole is about 1000 seconds at 1 mK in Cu and this rises in inverse proportion to the temperature. Thus you can see at once that the time to cool the metal and any other material is long, of the order of a day.

By such means the nuclear spin systems have been cooled to temperatures of the order of μK and lower, while an external thermometer attached to the metallic copper has certainly reached 7 μK.

Now that we are able to reach such very low temperatures, how are they measured?

The measurement of low temperatures

The general recipe for measuring low temperatures is: measure the temperature by the same means as you used to produce it. Thus where gases are used, use the gas scale until the pressures become too low to measure accurately (see Table 11). For temperatures reached by magnetic cooling, use the paramagnetism to measure the temperature. We now see how this is done.

First we consider how it is done in practice and then turn to the problem of ensuring that the practical thermometer does indeed measure absolute temperature.

For temperatures achieved by nuclear demagnetisation, pure platinum is commonly used as the thermometer and its nuclear magnetic susceptibility is measured by nuclear magnetic resonance. The susceptibility χ accurately obeys Curie's law in the required range of μK, i.e. $\chi = c/T$ where c is called the Curie constant. By calibrating the thermometer at one known high temperature the low temperature can then be found.

How do we know that Curie's law is obeyed? To be sure of this, we must first determine the absolute temperature in the new realm of temperature. To do this at the unknown low temperature, we define an *empirical* temperature scale by assuming that Curie's law still holds for our thermometric material. The temperature T_1^* on this scale is by definition:

$$T_1^* = \frac{c}{\chi_1}$$

We can now use this empirical scale to find the absolute temperature. Call the starting temperature for the adiabatic demagnetisation T_i; at this temperature we can find the entropy of the coolant as a function of the applied magnetic field B either by calculation from the Curie constant or by direct measurement of the heat evolved on magnetisation.

Now suppose we carry out a sequence of adiabatic demagnetisations from T_i at known starting fields corresponding to the points B, B', B'' etc. in Figure 21a. We measure T^* at the final temperature in zero field corresponding to the points C, C', C'' etc., which have, of course, the same entropy as B, B', B'',... respectively. Since we know the starting entropies we have now established the entropy in zero field as a function of T^*.

We now cool the coolant to its lowest temperature and add to it known small amounts of heat δq (all in zero field) to take it through the states C, C', C''... measuring T^* after each heating. (This process may not be entirely straightforward if the thermal conductivity of the coolant is low.) In this way we determine δq for the transitions from C to C' etc., for which we already know the corresponding values of δS.

We now make use of the thermodynamic relation, which defines absolute temperature:

$$dq = T\,dS$$

Therefore:

$$T \simeq \frac{\delta q(T^*)}{\delta S(T^*)}$$

By suitable interpolation and density of measurement we can make this as exact as we need and so find the absolute temperature T.

In fact this method of finding absolute temperature does not depend on measuring T^* from the susceptibility and can obviously be used with any secondary thermometer that is sensitive and reproducible in the temperature range of interest. Indeed these procedures are quite general. If there are interesting phenomena at still lower temperatures these will be associated with entropy changes and these entropy changes will in turn supply both the means to reach these new temperatures and the means of measuring them.

Exercises

Q1   Sketch on an entropy–temperature diagram the following sequence of quasi-static changes to 1 mole of He^4:

(a) The He starts as liquid at 1 atmosphere pressure at its normal boiling point 4.2 K; it is completely changed into vapour at the same temperature and pressure.
(b) The vapour is then raised in temperature to 10 K at atmospheric pressure.
(c) The gas at 10 K and 1 atmosphere pressure is then compressed isothermally to 100 atmospheres.
(d) The helium then performs an adiabatic reversible expansion to a pressure of 1 atmosphere. It reaches a temperature of 4.2 K and forms a mixture of liquid and vapour at the normal boiling point.

Calculate the entropy changes of the He^4 in each stage, given that the latent heat of vaporisation of He^4 is 93 J mole^{-1} at 4.2 K, that the molar heat capacity of gaseous He^4 at constant pressure is $5R/2$ and that the vapour and gaseous He^4 can be treated as an ideal gas. Hence calculate the fraction of liquid formed at the end of step (d). (The gas constant R has the value 8.3 J K^{-1} mole^{-1}.) A variant of these processes is used in a simple liquefier for helium, the Simon liquefier.

Q2   An object is to be cooled by using a Carnot engine working backwards as a refrigerator. A heat bath at a fixed temperature T_1 (e.g. room temperature) acts as the heat sink in which all the heat from the engine is deposited. Work W is done on the engine to extract heat from the object to be cooled. All the processes are ideally reversible and no other work or heat enters or leaves. The change in entropy of the object in going from T_1 to its final temperature is ΔS and that of its internal energy is ΔU. Show that the work W required (which is, of course, the minimum possible) is $\Delta U - T_1 \Delta S$. (This problem is almost the inverse of problem 3, Chapter 4.) The object to be cooled could be a mass of gas and the final state, if the temperature is appropriate, could be a liquid. The result would then give the minimum work required to liquefy the gas from the given starting conditions.

Q3 In the Linde method of liquefying a gas, the gas is compressed at room temperature from pressure p_1 (usually atmospheric) to the input pressure p_2 and then passed through a suitable Joule–Thomson valve. The cooling produced is passed on to the incoming gas by a heat exchanger. In both the heat exchanger and the valve, the enthalpy H of the gas is conserved. If H_1 is the enthalpy per unit mass of the gas going in at room temperature and if a fraction ε is liquefied when steady conditions have been reached, show that:

$$\varepsilon = (H_2 - H_1)/(H_2 - H_3)$$

where H_2 is the enthalpy per unit mass of the gas leaving the liquefier and H_3 is the enthalpy per unit mass of the liquid. Show also that under ideal conditions the work done per unit mass of liquid is then:

$$W = (RT/\varepsilon) \ln (p_2/p_1)$$

Assume that the heat exchanger is perfect so that the gas comes out at atmospheric pressure and room temperature T and that the gas at room temperature can be treated as ideal. (R here is the gas constant appropriate to the unit of mass chosen.)

16

The third law of thermodynamics and the unattainability of absolute zero

The first law of thermodynamics can be stated either as a positive assertion

$$\Delta U = Q + W$$

or in a negative form as 'it is impossible to make a perpetual motion machine which will produce work without consuming an equivalent amount of energy'.

The second law of thermodynamics can also be stated in a positive form, namely, that there exists a function of state called the entropy S defined by the relation

$$dS = \left(\frac{\text{d}q}{T} \right)_{rev}$$

Alternatively, it can be stated in the negative forms used by Clausius and Kelvin which have already been quoted.

The third law of thermodynamics can likewise be stated in either a positive or a negative form. We have already seen how the positive formulation can be made general and unambiguous. The negative form of the law states that 'it is impossible to reach the absolute zero of temperature by any finite number of processes'. Let us now briefly look at the third law from this point of view.

It is clear that attempts to reach low temperatures will be most efficient if they are carried out by reversible processes. Moreover, these processes should also be adiabatic since any heat entering the system to be cooled will impede our attempts to cool it. It is therefore clear that the best method of producing a lower temperature is by means of adiabatic reversible processes, i.e. processes at constant entropy.

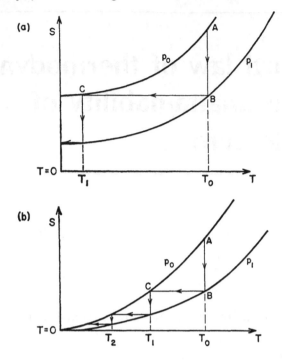

Figure 22 (a) Entropy–temperature diagram of a substance *not* obeying the third law. The absolute zero is accessible. (b) Entropy–temperature diagram of a substance obeying the third law. The absolute zero is unattainable.

If we look at the entropy-versus-temperature diagram of the substance with which we are trying to produce a low temperature, we can now see how the behaviour of the entropy at low temperatures is related to the accessibility of the absolute zero. This is illustrated in Figure 22 which shows the entropy of the substance at various constant values of the controlling parameter. In a gas this parameter might be the volume (or pressure); in the magnetic solid it would be the applied magnetic field. The figure shows two possibilities. In the first the entropy differences between different states of the system do not disappear as T tends to zero (in this example, therefore, the third law does not prevail). In the second, the entropy differences do disappear and the entropy of all states of the system tends to zero as the absolute zero is approached.

Imagine that T_0 is the temperature from which we start. Our first procedure is to reduce the entropy of the substance at T_0 by altering the controlling parameter from p_0 to p_1. The entropy thus changes from A to B. We then make the substance undergo an adiabatic reversible change from state B to state C (by altering the controlling parameter back to p_0) so that it cools from T_0 to T_1. Having now achieved the temperature T_1 we can in principle cool another body to this temperature and start the whole process from T_1. We thus make a reversible adiabatic change from T_1 and in this way reach the temperature T_2 and so on. In the first figure it would be quite possible by this process to reach the absolute zero, but in the second figure the converging of the different entropy curves makes this impossible except as a limiting process with an infinite number of steps.

In this discussion we have assumed that reversible changes are possible down to the lowest temperatures. If, however, such changes become impossible, this does not mean that it is easier to reach the absolute zero, but on the contrary it is even harder, since these irreversible effects will impede further progress towards low temperatures. The negative form of the third law, then, although on the whole not so immediately useful as the positive form, is quite general. As Professor Simon remarked: 'If someone finds a violation of the third law, all low temperature physicists will rejoice because we can then reach the absolute zero.'

The laws of thermodynamics

1st Law: You can't win, you can only break even.
2nd Law: You can break even only at the absolute zero.
3rd Law: You cannot reach absolute zero.

Conclusion: You can neither win nor break even.

(As quoted in *The American Scientist*, March 1964, p. 40A)

Notes

Part one

1   See e.g. *The Discovery of Specific and Latent Heats* by McKie and Heathcote, Edward Arnold & Co., London 1935.
2   *Lectures on the Elements of Chemistry*, Edinburgh 1803, 2 vols. Edited by Robison, vol. 1, p. 506.
3   *Philosophical Transactions of the Royal Society* 1798, p. 80.
4   *Philosophical Transactions of the Royal Society* 1850, **140**, part 1, p. 61.
5   See M. Born, *Natural Philosophy of Cause and Chance*, Clarendon Press, 1949.
6   For a discussion of the work of magnetisation and further references, see, for example, V. Heine, *Proceedings of the Cambridge Philosophical Society*, **52**, 546 (1956).
7   I have used W or Q to denote a large quantity of work or heat; w or q to denote a small quantity of work or heat; and $đw$ or $đq$ to denote the corresponding infinitesimal quantities.
8   Translations made from 1824 edition as published by La Librairie Scientifique et Technique, A. Blanchard, Paris 1953.
9   In the Clausius statement of the second law there is a reference to 'higher' and 'lower' temperatures. In this context we define the higher temperature as that from which the heat would flow spontaneously and the lower temperature as that to which the heat would flow. The heat influx can be recognised by methods based on the first law. Thus we do not rely on the gas scale or any other arbitrary scale of temperature to define higher and lower temperatures here.
10   William Thomson, *Mathematical and Physical Papers*, vol. 1, Cambridge University Press, 1882, pp. 179 and 181.
11   *Mathematical and Physical Papers*, vol. I, pp. 100 and 235.
12   *Annalen der Physik*, **93**, 481 (1854).
13   *Annalen der Physik*, **125**, 353 (1865).

14 The following is a very brief introduction to systems of variable mass, often called 'open' systems. Suppose that the number of moles N of substance in our system increases by dN at constant T and p. Let the internal energy, entropy and volume of the N moles be U, S and V while the corresponding quantities referred to one mole are u, s and v so that:

$$u = U/N, \quad s = S/N \quad \text{and} \quad v = V/N \tag{A1}$$

Changes in the quantities u, s and v, which refer to fixed mass, must therefore be related as before by the standard equation:

$$du = T\,ds - p\,dv \tag{A2}$$

But from equation (A1) we see that:

$$du = dU/N - U\,dN/N^2; \quad ds = dS/N - S\,dN/N^2; \quad dv = dV/N - V\,dN/N^2 \tag{A3}$$

If we now put (A3) in (A2) and rearrange we find:

$$dU = T\,dS - p\,dV + (U - TS + pV)\,dN/N \tag{A4}$$

The coefficient of dN is the Gibbs free energy per mole, which for a one component system is the chemical potential μ. Equation (A4) then becomes:

$$dU = T\,dS - p\,dV + \mu\,dN \tag{A5}$$

Thus, for 'open' systems the appropriate variables for the internal energy are $U(S, V, N)$. If one component is present in more than one phase, equilibrium occurs (i.e. the numbers of molecules of that component in any phase cease to change) when its chemical potential is the same in all the phases; for example, electrons in metals or semiconductors, joined together and in equilibrium, must have the same chemical potential throughout. When there is more than one component, equation (A5) becomes:

$$dU = T\,dS - p\,dV + \Sigma\mu_i\,dN_i \tag{A6}$$

where dN_i and μ_i refer to the ith component and the sum is over all components. The μ_i are now the partial Gibbs free energies.

15 To encompass all the essential features of equilibrium thermodynamics we should have to include the inequalities discussed on page 62. These are important in discussing the *stability* of thermodynamic systems.

16 See Thomson's *Mathematical and Physical Papers*, vol. I, p. 333.

Part two

1 An account of Boltzmann's work which I have found very helpful is given by R. Dugas in *La Théorie Physique au sens de Boltzmann et ses prolongements modernes*, Neuchatel-Suisse, Editions du Griffon, 1959. See also L. Boltzmann, *Lectures on Gas Theory*, translated by S. G. Brush, University of California Press, 1964.

2 *Wiener Berichte*, **76**, 373–435 (1877).

3 For a proof, see e.g. G. S. Rushbrooke, *Statistical Mechanics*, Clarendon Press, Oxford, 1948, p. 322.

4   In this Boltzmann was guided by Liouville's theorem in classical mechanics. See *Lectures on Gas Theory*, pp. 274–290 and also p. 443.

5   To obtain the result, we must also invoke a relationship between pressure and energy density.

6   At first sight this is surprising. For a more detailed discussion see Rushbrooke, *Statistical Mechanics*, p. 27.

7   Max Planck, *The Theory of Heat Radiation*, Dover Edition, 1959, p. 118.

8   A. E. Guggenheim, *Research*, **2**, 450 (1949).

9   A. Einstein, *Annalen der Physik*, **22**, 180 (1907).

10 The question of Gibbs' paradox arises at this point. See *The Scientific Papers of J. Willard Gibbs*, Dover Edition, vol. I, 1961, p. 166.

11 E_0 can be deduced from equation (142), p. 121. Notice, however, that in calculating E_0 here I have assumed one particle only per translational energy level. For particles of spin $\frac{1}{2}$ (e.g. electrons) there are two spin states associated with each translational level and these, in the absence of a magnetic field, would in general have the same energy. For particles of different spin there would be different numbers of spin states. This does not alter the argument but it does alter the numerical coefficients involved.

12 F. London, *Superfluids*, vol. 2, John Wiley and Sons, Inc., New York 1954.

13 This discussion is based on an article in *Physics World*, August 1995 by D. Meacher and P. Ruprecht. For a discussion of the conditions under which a composite particle is a boson, see *Elements of Statistical Mechanics* by ter Haar, Butterworth, 1995, Chapter 4.9.

14 J. C. Maxwell, *Theory of Heat*, Longman, Green and Co., London 1883, p. 328.

15 L. Brillouin, *Science and Information Theory*, Academic Press Inc., New York 1956.

16 L. Boltzmann, *Lectures on Gas Theory*, p. 444.

Part three

1   M. and B. Ruhemann, *Low Temperature Physics*, Cambridge University Press, 1937.

2   K. Clusius and A. Frank, *Zeitschrift für Electrochemie*, **49**, 308 (1943).

3   See Sir Francis Simon, *Year Book of the Physical Society*, London 1956, p. 1.

4   A fascinating account of the early history of low temperature physics is given by the Ruhemanns in the first few chapters of their book quoted in (1) above.

Answers to exercises

Chapter 2

A1 A resistor is immersed in a thermally insulated water bath with no electric current and no water flow. This, when all changes cease, is thermal equilibrium. If a constant electric current flows and water flows through the bath to carry away the heat flowing from the resistor so that all temperatures are unchanging, we have a steady state.

A2 $V^{-1}(\partial V/\partial T)_p$; $-V(\partial p/\partial V)_T$; $-V^{-1}(\partial V/\partial p)_T$; $(\partial M/\partial H)_T$.

A3 $W_{\text{on}} = -\int p\,\mathrm{d}V$. But $\beta_T = -V^{-1}(\partial V/\partial p)_T$ so $W = V\beta_T\int_{p_i}^{p_f}p\,\mathrm{d}p = V\beta_T(p_f^2 - p_i^2)/2 \approx 110\,\mathrm{J}$.

A4 Work done at input in 1000 s is $1000 \times 2 \times 10^6\,\mathrm{J} = 2 \times 10^9\,\mathrm{J}$. Work given out at the output is 1/2 this. The net work on the system is thus: $10^9\,\mathrm{J}$.

Chapter 3

A1 No. Yes.

A2 In the change $W = 0$ (expansion into a vacuum) and $Q = 0$ (thermal insulation). Therefore $\Delta U = W + Q = 0$.

A3 Since the isothermal and adiabatic compressibilities are the same to our degree of approximation, the work done here is the same as in example Q3 of Chapter 2, namely 110 J. Since here the process is adiabatic $\Delta U = W = 110\,\mathrm{J}$. Thus, the internal energy increases by this amount. In this adiabatic process, however, in contrast to the isothermal process, the temperature of the lead rises. (See example Q4 of Chapter 5.).

A4 For the path ACB, we have (using $\Delta U = W + Q$) $U_B - U_A = -20 + 50 = 30\,\mathrm{J}$.

(a) For path ADB, W is 5 times that along ACB, so $W = -100$ J. Since $\Delta U = 30$ J,
   $Q = 130$ J (in).
(b) For path BA, $W = +25$ J, $\Delta U = -30$ J, so $Q = -55$ J (out).
(c) For path AD, $\Delta U = U_D - U_A = 25$ J, $W = 0$ and so $Q = 25$ J (in).
   For path DB, $\Delta U = U_B - U_D = U_B - U_A + U_A - U_D = 30 - 25 = 5$ J;
   $W = -100$ J so $Q = 105$ J (in).

A5 $\Delta U = Q + W$, where here $W = -p\Delta V = 100 \times 10^5 \times 10^{-1} \times 10^{-3} = 1000$ J (on).
$Q = -3.36 \times 10^5$ J (out). So $\Delta U = -3.35 \times 10^5$ J (a reduction).

A6 We have: $đq = dU + p\,dV$ and here $dU = C_V\,d\theta$. Thus $C_p = đq/d\theta|_{P=\text{const}} = C_V + p(\partial V/\partial\theta)_p$. Since for 1 mole of ideal gas $pV = R\theta$, $p(\partial V/\partial\theta)_p = R$. Finally, therefore, $C_p = C_V + R$.

A7 With $đq = 0$ and $dU = C_V\,d\theta$ for an ideal gas, we have $C_V\,d\theta = -p\,dV$. For 1 mole, $pV = R\theta$ or $p\,dV + V\,dp = R\,d\theta$. We now substitute $d\theta = (p\,dV + V\,dp)/R$ and find: $(C_V + R)p\,dV + C_V V\,dp = 0$. With $C_V + R = C_p$, we then get: $(C_p\,dV/V) + (C_V\,dp/p) = 0$. Integrate to get: $C_p \ln V + C_V \ln p = \text{const.}$ or $pV^\gamma = \text{const.}$ with $\gamma = C_p/C_V$. Now put $p = R\theta/V$ and rearrange to give: $\theta V^{\gamma-1} = \text{const.}$

Chapter 4

A1 As we saw in example Q7 of Chapter 3, $TV^{\gamma-1} = \text{constant}$ in the reversible adiabatic change of an ideal gas. Thus for such a change from T_1, V_1 to T_2, V_2, $T_1/T_2 = (V_2/V_1)^{\gamma-1}$. Thus if $T_1 = 300$ K, $T_2 = 600$ K and $V_1 = 4$ l, $V_2 = 4 \times (1/2)^2 = 1$ l. At the other end of the 600 K isotherm, $V_1 = 1$ l and so $V_2 = 1/4$ l. For the corresponding pressures, we then have 8 atm and 32 atm respectively. At 300 K, the other extreme pressure is 4 atm. At the high temperature T_H, the heat absorbed is $nRT_H \int p\,dV = nRT_H \ln 4$. At the low temperature T_L, the heat given out is $nRT_L \ln 4$. From $pV = nRT$, we find that $nRT_H = 8$ l atm $\approx 8 \times 10^2$ J and so the heat absorbed at 600 K $\approx 1600 \ln 2$ J. The heat given out at 300 K is thus $800 \ln 2$ J. The thermodynamic efficiency is 0.5.

A2 The heat to be supplied to the house is at a rate of $Q_1 = 10$ kW at 300 K. If the work rate is W, we must have $W/Q_1 = (T_1 - T_2)/T_1 = 1/10$. Thus $W = 1$ kW.

A3 Let the temperatures of the two bodies at some intermediate stage be t_1 and t_2. In an infinitesimal process, let the engine receive heat $-C\,dt_1$ from body 1 and lose heat $C\,dt_2$ to body 2. For a Carnot engine we must then have $-C\,dt_1/t_1 = C\,dt_2/t_2$. The limits on the left are T_1 to T_F and on the right T_2 to T_F. Integrate to get $\ln T_F/T_1 = \ln T_2/T_F$ or $T_F = (T_1 T_2)^{1/2}$. The net change in internal energy of the two bodies is $C(T_F - T_2 + T_F - T_1) = 2C[T_F - (T_1 + T_2)/2]$. If, however, the bodies had exchanged energy purely by heat conduction, there would have been no net change in internal energy; one body would have gained what the other lost. The difference is accounted for by the work produced, which is thus:

$$W = 2C[(T_1 + T_2)/2 - (T_1 T_2)^{1/2}].$$

A4 We are given that $V^{-1}(dV/dT)_p = 3.66 \times 10^{-3}$. We also know that on the absolute scale, at constant pressure, $T \propto V$ or $\partial T/T = \partial V/V$. Therefore, $V^{-1}(\partial V/\partial T)_p = 1/T$. At the ice point T is thus $1/3.66 \times 10^{-3} = 273$ K.

Chapter 5

A1 From 20°C to 0°C, $\Delta S = 4200 \ln (273/293) = -297 \text{ J K}^{-1}$
At 0°C $\Delta S = -3.36 \times 10^5/273 = -1231 \text{ J K}^{-1}$
From 0°C to -10°C, $\Delta S = 2100 \ln (263/273) = -78 \text{ J K}^{-1}$
ΔS_{total} $= -1606 \text{ J K}^{-1}$

Use $dp/dT = \Delta S/\Delta V$.
For the transition ice to water, $\Delta S = 1231 \text{ J K}^{-1}$ per kg and $\Delta V = -0.1 \times 10^{-3} \text{ m}^3$ per kg. So $dp/dT = -1.23 \times 10^7 \text{ Pa K}^{-1}$.

A2 Use $dp/dT = \Delta S/\Delta V$, where p is the vapour pressure at T, ΔS is the entropy of vaporisation and ΔV the volume change on vaporisation. Here $\Delta S = 10R$ according to Trouton's rule. $\Delta V = V_v - V_l$, where v refers to the vapour and l to the liquid. We ignore V_l and treat V_v as if the vapour obeyed $pV_v = RT$. Substitute and we get: $dp/dT = 10p/T$, which we can integrate to give: $\ln p = 10 \ln T + \text{const}$.

A3 $(\partial S/\partial p)_T = -(\partial V/\partial T)_p = -10^{-3} \times 8 \times 10^{-5} \text{ J K}^{-1} \text{ Pa}^{-1}$. Thus, in increasing the pressure to 1000 atm or 10^8 Pa, $\Delta S = -8 \times 10^{-8} \times 10^8 = -8 \text{ J K}^{-1}$. So the heat given out is 2400 J. $\Delta U = Q + W$ and from Q3 of Chapter 2, $W = 110 \text{ J}$. So $\Delta U = -2400 + 110 = -2290 \text{ J}$.

A4 $(\partial T/\partial p)_S = (\partial V/\partial S)_p = (\partial V/\partial T)_p (\partial T/\partial S)_p = V[V^{-1}(\partial V/\partial T)_p] T/T(\partial S/\partial T)_p = TV\alpha/C_P$, where α is the thermal expansion coefficient.
So $\Delta T = TV\alpha\Delta p/C_p = 300 \times 10^{-3} \times 8 \times 10^{-5} \times 10^8/25 \times (10^{-3}/18.3 \times 10^{-6}) = 1.8 \text{ K}$.

A5 (1) In going from 100 to 400 K at constant volume, $Q = 450R$ and the pressure becomes 4×10^5 Pa. $\Delta S = (3R/2) \ln 4 = 3R \ln 2$. In going from 4×10^5 to 8×10^5 Pa isothermally, $Q = -400R \ln 2$ and $\Delta S = -R \ln 2$. Thus the total Q is $450R - 400R \ln 2 = 172.7R$ for this path. $\Delta S = 2R \ln 2$.
(2) In going from 100 to 400 K at constant pressure, $Q = 750R$ and $\Delta S = (5R/2) \ln 4 = 5R \ln 2$. In going from 10^5 Pa to 8×10^5 Pa isothermally, $Q = -400R \ln 8$ and $\Delta S = -R \ln 8 = -3R \ln 2$. Thus the total Q is $750R - 1200R \ln 2 = -81.8R$ for this path (of opposite sign even!). $\Delta S = 2R \ln 2$ as before.

Chapter 6

A1 (a) No. (b) Yes. (c) Yes. (d) $\Delta U = Q + W$ (per second). $\Delta U = 0$ as none of its parameters change. So $W = -Q$. The work done electrically appears as a heat flow. (e) No. (f) Yes, the entropy of the bath increases. (g) The entropy increases as $Q/T = 33.3 \text{ JK}^{-1}\text{s}^{-1}$.

A2 (a) Initial entropy $S_i = R \ln 5 + \text{const}$. Final entropy $S_f = R \ln 10 + \text{same const}$. $\Delta S = R \ln 2$. No change in the entropy of surroundings. (b) $\Delta S = R \ln 2$ for oxygen; $\Delta S = -R \ln 2$ for the bath.

Chapter 7

A1

n_0	n_1	n_2	n_3		P
1	2		1		3
2	1	1	1		6
3	3			1	

No. 3 is the least and no. 2 the most probable. The average distribution is $n_0 = (2 \times 3 + 1 \times 6)/10 = 1.2$; $n_1 = (1 \times 6 + 3 \times 1)/10 = 0.9$; $n_2 = (1 \times 6)/10 = 0.6$; $n_3 = (1 \times 3)/10 = 0.3$.

A2 When rearranged, equation (c) becomes:

$$x(1 + \lambda) + y(1 - \lambda) - (3 + \lambda) = 0$$

With $\lambda = -1$, we find $y = 1$; with $\lambda = +1$, we find $x = 2$.

Chapter 8

A1 We would expect appreciable ionisation when kT is about 13.6 eV. Thus T is about $13.6 \times 1.6 \times 10^{-19}/1.4 \times 10^{-23} \simeq 10^5$ K.

Chapter 9

A1 Energy levels are 0 and $2\mu B$ so $n_\downarrow/N = \exp(-2\mu B/kT)/[\exp(-2\mu B/kT) + 1]$ $n_\uparrow/N = 1/[\exp(-2\mu B/kT) + 1]$. Thus $(n_\uparrow - n_\downarrow)/N = [1 - \exp(-2\mu B/kT)]/[\exp(-2\mu B/kT) + 1] \simeq (2\mu B/kT) = 1.41 \times 10^{-26}/1.36 \times 10^{-23} \times 300 \approx 3 \times 10^{-6}$. This small difference in occupation numbers is sufficient to allow an absorption of radio frequency energy when the quantum of R.F. energy matches the energy separation of the two levels.

Chapter 10

A1 Partition function is: $\exp(-\varepsilon_1/kT) + 2\exp(-\varepsilon_2/kT)$. At high T, S tends to $R \ln 3$ and at low T to zero. T is high or low according as $kT \gg$ or $\ll \varepsilon_2 - \varepsilon_1$.

A2 (a) An energy of 10^{-11} eV is equivalent to a temperature of $10^{-11} \times 1.6 \times 10^{-19}/1.4 \times 10^{-23} \simeq 10^{-7}$ K. So the nuclear spin entropy for a spin 1/2 nucleus at 1 K will be $R \ln 2$. The contribution to the heat capacity from this source will be negligible. There will be a small T^3 contribution from the lattice vibrations (see p. 108). (b) At 300 K the heat capacity of the lattice will have achieved its classical value of $3R$ per mole.

A3 As $T \to$ infinity, all the exponential Boltzmann factors tend to unity so that $n_0 = n_1 = n_2 = N/3$ where N is the total number of particles. Thus the average energy per particle is: $[(\varepsilon_0 + \varepsilon_1 + \varepsilon_2)N/3]/N = (\varepsilon_0 + \varepsilon_1 + \varepsilon_2)/3$. To take more energy some particles must be promoted from the ground state.

Chapter 11

A1 To find the number we seek, we first imagine that all the ○'s and bars are different. Then the number of distinguishable arrangements is $(n_i + w_i - 1)!$. But since, in fact, interchanging the ○'s among themselves and the bars among themselves produces nothing new, the number of distinguishable ways of distributing n_i indistinguishable objects in w_i boxes is:

$$p_i = (n_i + w_i - 1)!/n_i!(w_i - 1)!$$

A2 (a) The final pressure is $p_3 = 2$ atmospheres. The initial entropy is $C - R\ln p_1 - 3R\ln p_2$ where C is a constant; the final entropy is $C - 4R\ln p_3$. So the increase in entropy $\Delta S = R[\ln 1 + 3\ln 3 - 4\ln 2] = R\ln(3^3/2^4) = R\ln(27/16)$. (b) For each gas the pressure is now halved so the increase is here $\Delta S = R\ln 2 + 3R\ln 2 = R\ln 16$.

A3 We have to estimate the moment of inertia of the single atom and of the diatomic molecule and see what rotational energies are involved. $I \approx mr^2$. For the nucleus we get $I \approx 10^{-26} \times 10^{-34} \approx 10^{-60}$ kg m^2 and so the separation of the first two rotational energy levels is of order 10^{-8} J; this is equivalent to a temperature of 10^{15} K! It will not be excited. On a classical picture the moment of inertia of an electron about the nucleus is of order 10^{-46} with an excitation temperature of order 10^5 K, too high to contribute. For the diatomic molecule, $I \approx 10^{-46}$ with an excitation temperature of order 10 K, which is obviously within the range of normal temperatures. This, however applies only to the two principal axes of rotation normal to the axis of the molecule: if the axis of rotation coincides with that of the molecule the value of I is comparable with the nuclear moment of inertia and this motion is not excited.

Chapter 12

A1 (a) Each atom can occupy each state and contribute to a new microstate. There are thus $3^3 = 27$ distinguishable microstates. (b) There is now only one microstate, with each quantum state singly occupied. (c) We use the expression derived in the text: $P = (n + w - 1)!/n!(w - 1)!$ with here $n = w = 3$. Thus there are 10 microstates.

A2 $Mv^2/2 = 3RT/2$. So $v \approx (3 \times 8.3 \times 10^{-7}/85 \times 10^{-3})^{1/2} \approx 5 \times 10^{-3}$ m s^{-1}.

A3 4.7 eV corresponds to a temperature $T_F = 4.7 \times 1.6 \times 10^{-19}/1.4 \times 10^{-23}$ K $\approx 5.4 \times 10^4$ K. The fraction of electrons excited at 300 K is thus roughly $300/5.4 \times 10^4 \approx 6 \times 10^{-3}$. The energy carried by 1 mole of electrons at temperature T is $U \approx RT \times (T/T_F)$ and so the heat capacity $dU/dT \approx 2R(T/T_F) \approx 10^{-2}R$ at 300 K.

A4 For free particles of mass m, the Fermi energy $E_F \propto m^{-1}$ and so here the ratio is 2000 times higher for the electrons. If in a star a proton and an electron can combine to form a neutron, the energy of the star can be greatly reduced. The high kinetic energy of the electrons is replaced by the 2000-fold lower kinetic energy of the neutrons.

Chapter 13

A1 (a) The viscosity; the electrical or thermal conductivity; the diffusion coefficient. (b) The mean free path of gas molecules; the relaxation time of electrons in metals; the mobility of electrons or holes in semiconductors. (c) The electron mass; the moment of inertia of a gas molecule; vibrational or rotational energy levels of a molecule; the electronic energy states of atoms etc.

Chapter 15

A1 See figure for the corresponding points ABCDE.

(a) From A to B, $\Delta S = L_{vap}/T = 93/4.2 = 22\,\mathrm{J\,K^{-1}}$.

(b) From B to C, $\Delta S = (5R/2)\ln(10/4.2) = 20.8 \times 0.87 = 18\,\mathrm{J\,K^{-1}}$.

(c) From C to D, $\Delta S = -R\ln 100 = -38\,\mathrm{J\,K^{-1}}$.

(d) From D to E, $\Delta S = 0$.

At 4.2 K, the final entropy of the helium is $22 + 18 - 38 = 2\,\mathrm{J\,K^{-1}}$ above that at its starting point. This, as indicated on the diagram, is AE and so the point E corresponds

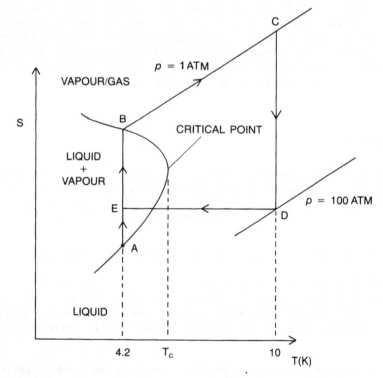

Figure A1 Entropy–temperature diagram of He^4 (schematic) near its normal boiling point. If the temperature scale is logarithmic and the entropy scale linear, the isobars of the ideal gas are straight lines. The points A, B, C, D and E refer to the stages in the problem.

190

to a fraction EA/BA of vapour. Thus the final state consists of 2/22 parts vapour and 20/22 parts liquid.

A2 If the engine cools the object by $-dT$ at temperature T the heat removed is $dq = -C\,dT$, where C is the heat capacity of the object at T. The work done is given by $dw/dq = (T_1 - T)/T = T_1/T - 1$. Thus $\int dw = -\int T_1(C\,dT/T) + \int C\,dT = \Delta U - T_1 \Delta S$.

A3 For unit mass the enthalpy entering the process is H_1. The enthalpy leaving the process is εH_3 for the liquid and $(1 - \varepsilon)H_2$ for the returning gas. So $H_1 = \varepsilon H_3 + (1 - \varepsilon)H_2$; i.e. $\varepsilon = (H_1 - H_2)/(H_3 - H_2)$. For unit mass of liquid, $1/\varepsilon$ is the mass of gas that must pass through the system. The work done to compress this from p_1 to p_2 is $(RT/\varepsilon) \ln (p_2/p_1)$.

191

Reading list

Classical thermodynamics

E. A. Guggenheim, *Thermodynamics*, North Holland Publishing Company, Amsterdam. (For reference.)

A. B. Pippard, *Classical Thermodynamics*, Cambridge University Press.

Mark W. Zemansky, *Heat and Thermodynamics*, McGraw Hill Book Company, New York and London. (Treats the first law by the Born approach; early editions shorter and better than the later ones.)

Statistical mechanics

G. S. Rushbrooke, *Statistical Mechanics*, Clarendon Press, Oxford. (Clear introductory treatment, yet with considerable detail.)

R. C. Tolman, *The Principles of Statistical Mechanics*, Oxford University Press. (Concerned with the fundamentals of the subject.)

F. Mandel, *Statistical Physics*, John Wiley and Sons, Chichester.

Tony Guenault, *Statistical Physics*, Chapman and Hall, London. (Informal, readable.)

Experimental

Guy K. White, *Experimental Techniques in Low Temperature Physics*, Oxford Science Publications. (Plenty of practical information and references to other sources.)

Index

195

and unattainability of absolute zero 177–9
Thompson, Benjamin (Count Rumford) 5–7
Thomson W. (Lord Kelvin) 33, 35–6, 140, 177
triple point 47, 52, 149, 168–9

unattainability of absolute zero 177–9

vapour pressure 168–70

velocity distribution in a gas 68–9, 123
work
  in irreversible processes 15–16
  in quasi-static processes 14–15
  in statistical mechanics 96
  of liquefaction 176, 191

zero point energy 80, 106, 149
zeroth law of thermodynamics 3, 13, 49

9 780748 405695

Milton Keynes UK
Ingram Content Group UK Ltd.
UKHW040058071024
449327UK00019B/638